The New Indicators of Well-Being and Development

Also by Jean Gadrey

NEW ECONOMY, NEW MYTH (ed.)
PRODUCTIVITY, INNOVATION AND KNOWLEDGE IN SERVICES
(ed. with Faïz Gallouj)

The New Indicators of Well-Being and Development

Jean Gadrey and Florence Jany-Catrice

First published 2006 by
PALGRAVE MACMILLAN
Houndmills, Basingstoke, Hampshire RG21 6XS and
175 Fifth Avenue, New York, N.Y. 10010
Companies and representatives throughout the world

PALGRAVE MACMILLAN is the global academic imprint of the Palgrave Macmillan division of St. Martin's Press, LLC and of Palgrave Macmillan Ltd. Macmillan® is a registered trademark in the United States, United Kingdom and other countries. Palgrave is a registered trademark in the European Union and other countries.

ISBN 13: 978–0–230–00500–6   hardback
ISBN 10: 0–230–00500–4   hardback

This book is printed on paper suitable for recycling and made from fully managed and sustained forest sources.

A catalogue record for this book is available from the British Library.

Library of Congress Cataloging-in-Publication Data
Gadrey, J.
     The new indicators of well-being and development / Jean Gadrey and Florence Jany-Catrice.
         p. cm.
     Translation of: Les nouveaux indicateurs de richesse.
     Includes bibliographical references and index.
     ISBN 0–230–00500–4
     1. Economic indicators. 2. Quality of life—Measurement.
     3. Sustainable development. 4. Environmental economics. I. Jany-Catrice, Florence. II. Title.
     HD75.G325 2006
     306.3'61–dc22   2006040029

10   9   8   7   6   5   4   3   2   1
15   14   13   12   11   10   09   08   07   06

Transferred to digital printing 2006

Contents

List of Tables

List of Figures

List of Acronyms and Abbreviations

Indicators

BIP 40:	Baromètre des Inégalités et de la Pauvreté (Barometer of Inequalities and Poverty) (Chapter II)
GDI:	Gender-Related Development Index (Chapter III)
GDP:	Gross Domestic Product (Chapter II)
GEM:	Gender Empowerment Measure (Chapter III)
GNP:	Gross National Product, now Gross National Income (SCN 93/SEC 95) (Chapter V)
GPI:	Genuine Progress Indicator (Chapter V)
HDI:	Human Development Index (Chapter III)
HPI:	Human Poverty Index (Chapter III)
IEWB:	Index of Economic Well-Being (Chapter VI)
ISEW:	Index of Sustainable Economic Welfare (Chapter V)
ISH:	Index of Social Health (Chapter III)
MEW:	Measure of Economic Welfare (Chapter IV)
PSI:	Personal Security Index (Chapter III)
SMEW:	Sustainable Measure of Economic Welfare (Chapter IV)
TDI:	Technical Development Indicator (Chapter III)

Organizations

CCSD:	Canadian Council on Social Development
CERC:	Centre d'Etude des Revenus et des Coûts
CGSDC:	Consultative Group on Sustainable Development Indicators
CSLS:	Center for the Study of Living Standards
INSEE:	Institut National de la Statistique et des Etudes Economiques (French National Institute for Statistics and Economic Studies)
NEF:	New Economic Foundation

NGO: non-governmental organization
OECD: Organization for Economic Co-operation and
        Development
RP: Redefining Progress
UNDP: United Nations Development Programme
WWF: World Wildlife Fund, renamed World Wide Fund for
        Nature in 1987

Others

ESA 95: European System of Accounts
GFCF: Gross Fixed Capital Formation
PPP: Purchasing Power Parities
R&D: Research and Development
SEEA: System of Economic and Environmental Accounts.
        Compiled since 2003 by the United Nations; see:
        http://unstats.un.org/unsd/envAccounting/seea.htm
SNA 93: System of National Accounts (UN, IMF, OECD, World
        Bank, Eurostat)

Acknowledgement

This book has benefited from collaboration with various people, among them Thierry Ribault and Bruno Boidin, and from the support of Dominique Méda. We sincerely thank our translator, Andrew Wilson, for his most important contribution to the adaptation of the content of the French edition, and Janey Fisher for her valuable suggestions at the final editing stage.

Introduction

When drawing up their overall judgements on progress or expressing their concerns about the future, developed societies still rely heavily – perhaps even more than they used to – on the main indicator of economic growth, which measures variations in gross domestic product (GDP) or its variants and remains the cornerstone of national accounting. Of course in the public debate and in the media, other major indicators regularly attract attention, particularly the unemployment rate, the inflation rate and, more recently, the main stock-exchange indices. However, the level of GDP and growth are still the main symbols of success. Since the 1970s, the dominance of these criteria has attracted criticism, some of it from economists but, more frequently, from various 'anti-establishment' actors, some of whose protests have had a social slant (growth is not necessarily to be equated with social progress) while others have been more concerned with environmental issues (growth can destroy natural resources). However, these criticisms have had little impact until now, at least as far as the institutionalization of alternative indicators is concerned.

There are several reasons for the ineffectualness of this criticism. The main one is that, even though it is true that growth is not the answer to all problems, many people believe, quite rightly, that it can open up room for manoeuvre and improve certain aspects of daily life, employment, and so on. In the short and medium term, therefore, growth is regarded positively by large sections of the population. This widespread acceptance of the benefits of growth is, incidentally, further reinforced by the fact that they

1

are encapsulated in one final figure (the rate of growth achieved or hoped for) that ignores the qualitative aspect of growth (what has improved?). Great care is taken to avoid specifying, for example, who has received what (the 'distribution of value added'). And even less mention is made of certain problems of measurement which are, nevertheless, quite formidable and would undoubtedly undermine the religion of the growth rate if they were ever to become the focus of public attention (Gadrey 2002).

However, there are other explanations for the relatively little impact these criticisms have had. One of them is the weakness of the alternative indicators that have been suggested in the past; this weakness is due in part to theoretical and methodological shortcomings and part to their inability to make sense in the public sphere. Nevertheless, it would seem that recent years have seen genuine advances in the search for indicators that are both theoretically 'defensible' and capable of reflecting an extended notion of wealth and well-being. We cannot of course say for certain today whether these innovations will gain widespread acceptance and acquire a social and technical status comparable to that of growth, but the fact is that they are proliferating and, in some cases, converging. The present book is given over to an international comparative assessment of these innovations. It is concerned with research studies, surveys and statistical investigations whose principal objective is to develop new *synthetic* indicators, either for the purpose of international comparisons or in order to measure variations over time. What these indicators have in common is the explicit desire of their creators and advocates to provide a counterbalance to purely economic evaluations of progress. Thus in virtually all cases, we are dealing with 'socio-economic' indicators, in the sense that they attempt to take into account not only economic criteria but also human, social and, in many instances, environmental criteria. They can be described as 'alternative', not only because they seek to offer an alternative to the traditional and dominant indicators of economic growth or consumption but also because they complement those indicators without, in many cases, abandoning them altogether as elements in the overall assessment they provide.

Our main concern is with the major studies carried out in the 1990s. This choice can be easily justified by simply examining what was produced and diffused over that period and comparing it to the comparatively few emanating from the decade of the 1980s. What is striking is the rapid expansion of the battery of socio-economic indicators that have managed to become sufficiently well-known to be incorporated into the public debate in various forms at both national and international level (albeit with limited political impact to date, to judge from the slow pace of change in public regulation and policy). The criteria we adopted in order to identify the 'major' studies is *international renown, in both academia and the media*. Several of these studies have also been taken up and commented on in recent publications produced by major international organizations such as the OECD, the World Bank and the United Nations. Virtually all of them have been the subject of academic articles in reputable international journals. We proceeded to select those which, in our view, are the most influential today across the world – which excludes in particular the Kids Count Index, synthetic indicators of quality of life and Richard Estes' indicator of social progress (see Gadrey & Jany-Catrice 2003).

Also excluded from the book are those studies that do offer assessments or multidimensional 'dashboards' *but not for the purpose of developing a synthetic indicator*. Studies of this kind are included in the report from which this book is derived (Gadrey & Jany-Catrice 2003), as well as in the assessment subsequently carried out in Canada by Sharpe (2003), but we had to make certain choices. In our view, the best way of giving an idea of what alternative indicators might add to the public debate on wealth is to provide examples in which these indicators both complement and offer credible alternatives to GDP and growth. The synthetic indicators have this capability, more than complex 'dashboards' of multiple data. They can be a first 'focus of attention' comparable to GDP and growth.

Of course, the construction of synthetic indicators has always to be preceded by an assessment exercise, with the 'aggregation' process taking place subsequently; conversely, an assessment exercise can always be extended by the construction of synthetic indicators. Nevertheless, these two processes differ from each other, in terms both of the authors' intentions and the use to which they are put by policymakers and the media. In our view, they are complementary.

The attention paid to synthetic indicators, still known as aggregate indicators, has increased considerably in recent years, so much so indeed that the OECD itself, which was initially reluctant to go down this route, published a report in April 2002 that contained an international evaluation of predominantly 'green' synthetic indicators.[1]

Our first hypothesis is that, in some respects, the political and intellectual situation with regard to the need for new tools is reminiscent of that in the developed world following the depression of the 1930s and the Second World War. Of course, the current global social and environmental crises are different in nature to the situation following the Second World War, and the current challenges cannot be adequately addressed at national level alone, unlike that of post-war 'reconstruction', since they are essentially global issues. However, the situation in the Southern countries, the problems of inequality and exclusion in many of the Northern countries and the state of the natural environment throughout the whole world are dramatic and potentially explosive. From Seattle to Porto Alegre and Johannesburg, global protests are growing and are likely to intensify. A new value system and a new hierarchy of factors that 'count' are set to gain ground and impact on political discourse and decisions.

However – and this is our second hypothesis – the principal national and international statistical indicators, the ones that are most widely diffused, receive the greatest exposure in the media and are consequently the most influential, are forms of knowledge that depend on political conventions and value systems. They are partially autonomous forms, of course, since the controversies that have accompanied their creation and diffusion also have complex academic and technical dimensions, some of which are unconnected with the general representations of social progress. However, the societal indicators that take pride of place owe their supremacy, primarily, to the fact that they have been politically selected following national and international conflicts or power struggles, in which economists, statisticians and accountants themselves contribute, in their own specific ways, to one or other aspect of the political debate.

If our two hypotheses prove to be correct, the decades to come (and we should remember that it took two decades of intense debate to lay the foundations of the current national accounting systems) should see the emergence of various expectations that will require the national and international accounting systems of the twenty-first century to be both economic accounts that extend the current national accounting systems and accounts that reflect 'sustainable human development'. These two sets of accounts would be linked and placed on the same footing in technical, institutional and political terms as well as in media portrayals. Words have a greater or lesser ability to create alliances, depending on their previous political history. The combination of human development (in the very broad sense attributed to it by the UNDP) and sustainable development is probably what today best sums up the political objectives of the accounts of the future. It is they that most strongly link the urgent demands expressed by the Southern countries, the expectations of environmentalists (currently articulated by certain social categories in the Northern countries) and the return of 'social issues' (poverty, inequality, violence) to both Northern and Southern countries.

I
The Search for Alternative Indicators

1 Indicators, value systems and the formation of judgements

Clearly, the indicators used by nation states to depict their progress and their failures, both to their own populations and to the outside world, are influenced by important political and social issues. No less clearly, progress evaluations and political preferences are shaped by many other forces and sources of information, which in turn rely to varying degrees on reason and passion. We should at all costs, therefore, avoid turning social and environmental indicators into a religion that might replace that of economic growth. This would amount to nothing more than a fetishization of measurement tools and assuming that everything that counts can be counted. It is still the case, however, that certain general indicators play a tremendously important role in the public debate and that the importance of this role seems to have increased at a time when economic growth has begun to slow down and become uncertain, as if this uncertainty heightened awareness of what is at stake and further concentrated attention on a single indicator. One of our hypotheses is that the major social and economic indicators are not simply passive reflections of the phenomena they claim to summarize. Along with other elements of our informational environment, they also help to structure our cognitive frameworks, our vision of the world, our values and our judgements. Thus the de facto dominance some of them enjoy is not a neutral matter. The question of the indicators used to measure wealth is a matter for the population as a whole and requires that debates which are presently confined to a small circle of experts be opened up to a wider public.

We would go even further. Several of the synthetic indicators we will come across here are based on variants of the notion of collective 'well-being', assessed at the level of the nation state. There are well-established objections to this notion, which stress the impossibility of 'aggregating individual preferences' in order to arrive at a collective notion of 'well-being'. This impossibility is 'demonstrated' mathematically on the basis of very specific hypotheses, in which individuals are endowed with given preferences; they do not discuss them with each other with a view to changing and bringing them closer together, so that options are closed off. Instead, they vote for a number of fixed alternatives that are offered to them; they are not in a position to produce any alternative. On this basis, any synthetic indicator handed down from on high could be condemned because it imposes, in a more or less dictatorial fashion (dictatorship being the only way of overcoming the impossibilities in question, once unanimity is excluded), one vision of 'well-being', namely that of the index designers. However, it is possible to see things differently, as Amartya Sen (1999), among others, does. The choice of weightings for the different variables that make up an indicator, just like the choice of the variables 'that count', is a matter for public debate and careful scrutiny of the differing views about what should be included and what should count the most (see Box I.1). The aim of such scrutiny should be as much to clarify the areas of disagreement as to build agreements that are sufficiently extensive and durable to become established without having to be imposed. Thus an indicator's legitimacy is constructed at the same time as the 'conventions for assessing progress'. GDP is not immune to this social law (Gadrey 2003). In this respect, an indicator, whether synthetic or not, seems all the more likely to form the basis for a non-imposed, lasting convention the more transparent it is (with regard to the values it conveys and its criteria, sources and methods) and the more readily it lends itself to the development of variants that can be debated beyond the narrow circle of its designers. This is another way of thinking about the social dynamic around these indicators rather than focusing simply on the social problems they point to.

Like all other indicators, the alternative indicators that will be outlined here reflect value systems and notions of what constitutes a

Box I.1   Weights and conventions

In many cases, the alternative indicators are blamed because they use 'arbitrary' or 'subjective' weights to aggregate their component variables. For example, the UNDP's HDI (see chapter III) is the average of three indicators of health, education and income, respectively. This means that each is as important as the other in assessing human development. But some people might value health more than education. If so health should get a higher weighting. Young people might not value health as much as older people, because they take it for granted. For the same reasons, rich people might not value income as much as education. And so on. Certainly, the choice of the same weights for the three component indicators is 'arbitrary'. But that does not mean that these choices are meaningless or devoid of interest for international comparisons or for assessing progress in each country. If a synthetic index built as an average of components declines in one country, or if it is lower in a country than in another, the question will be raised: which components are concerned by a bad performance? The synthetic index appears to be a first 'focus of attention' which then leads to more complex multidimensional analyses. After all, the NYSE and the NASDAQ are also 'arbitrary' averages of component variables. They do not properly reflect the situation on any individual owner of stocks. Are they meaningless?

'good society'. They have no pretensions to neutrality. It is only by recognizing this fact that attempts can be made to classify them, to identify those that, in quantified assessments of progress, put the emphasis on the re-introduction of the social and human dimensions, and those that are geared more to environmental questions.

2 Typologizing the studies surveyed, with an emphasis on synthetic indicators

How can we set about this survey in a methodical and systematic way? Our solution is to use two main criteria: 'values', on the one hand, and 'method', on the other. After all, these alternative studies are driven by concerns that differ to some extent, although they frequently converge; in some cases, it is 'social' values (human and social development) that dominate, while in others it is environmental values. And they use two different types of method. Some are based on the average (weighted or otherwise) of heterogeneous indexes, without any attempt to monetize the constituent variables (that is, to express their value in monetary units), while others use a method of national accounting (i.e. monetization) which has been extended to include variables that are currently excluded. Only the very important studies by Osberg and Sharpe (see Chapter VI) combine these two methods in an original way.

It so happens, however, that, in reality, these two criteria overlap to a fairly large extent: the vast majority of indicators that emphasize the social and human dimensions use the first method, while the vast majority of synthetic indicators with a strong environmental focus use the second one. This is why Chapters II and III, which focus on the non-monetized indicators, are concerned with more social themes, while Chapters IV and V, which deal with extensions of the national accounting method, are concerned with indicators with a strong environmental focus.

The complementarity of synthetic indicators and multidimensional assessments

Without dwelling on the point, we should mention the problem of the adoption or rejection of synthetic indicators, which are generally pitted against multiple indicators in the form of 'assessments' or 'dashboards'. It seems to us that it is not a question of choosing between synthetic and multiple indicators but rather of developing the two options simultaneously since, given the appropriate degree of transparency, they can both enhance public debate, mutually reinforce each other and contribute to the individual and collective formation of judgements on progress. To cite just one example, Amartya Sen, the celebrated economist who was the inspiration for the UNDP's pioneering work on these matters, was initially not in favour of publishing a synthetic indicator (the HDI, or human development index), regarding it as too 'crude', whereas the available data, taken as a whole, were extremely rich. Sen retracted this view in 1999. Referring to his earlier debates

Box I.2   Other criteria for differentiating the indicators

The two main criteria we have used to differentiate the indicators and to structure this book are not the only ones that could be envisaged. For example, one important distinction that can be made is that between 'objective' and 'subjective' indicators. The latter are based on surveys of people's opinions and 'feelings' (trust, safety, etc.), whereas the former are based on data that do not explicitly contain any value judgements about the situation experienced. In reality, this distinction is not so simple. Many surveys that aim to produce 'objective' data are based, for example, on personal statements made by those surveyed, whose subjectivity is consequently involved to varying degrees. Even 'administrative data', including economic data, are generated by means of procedures that are always shaped by human decisions. This does not mean that the distinction between these two types of indicators is meaningless, but rather that the distinction between the two should be put into context. 'Subjective' indicators constructed with serious intent can be more significant and more robust than some so-called objective indicators, particularly when it comes to monitoring developments over time in a single country. Most of the indicators that have names such as 'life satisfaction', 'happiness' or 'subjective well-being' fall into the second category and are far from being without value, particularly when efforts are made to combine them with objective indicators. One good example is the Canadian 'personal security index', which will be presented in Chapter III.

with the other 'father' of the UNDP's reports and indicators, Mahbub ul Haq, Sen wrote:

> Mahbub got this exactly right, I have to admit, and I am very glad that we did not manage to deflect him from seeking a crude measure. By skilful use of the attractive power of the HDI, Mahbub got readers to take an involved interest in the large class

of systematic tables and detailed critical analyses presented in the *Human Development Report*. The crude index spoke loud and clear and received intelligent attention and through that vehicle the complex reality contained in the rest of the Report also found an interested audience.

(UNDP 1999)

3 A first indicator of growth

Our first observation, following the survey we carried out, was that there has been a veritable explosion in the number of major initiatives in the course of the 1990s, and more especially since 1995. There were two main driving forces: social issues (including the problems associated with human development and that of the quality of life) and environmental issues. The available stock of alternative macro-level socio-economic indicators (including structured dashboards of indicators) increased from practically zero during the 1980s to two in 1990 (the UNDP's HDI and the Kids Count Index, which is not discussed in this book), then to 12 in 1995 and to no fewer than 30 by 2001–02!

We are not suggesting that this exceptional increase will continue for very long. It has probably not finished; we seem to be at the beginning of the 'S-curve' for such innovations, with the likelihood that they will become more widely diffused in developed countries and then elsewhere and that a series of convergences, alliances and regroupings will take place.

Further information can be provided on the various elements in this rapid increase. In the early 2000s, 'non-monetized' indicators account for much the highest share of new indicators (18 out of the 29 surveyed). Some of the indicators in this group are synthetic, others are not and their construction is driven largely by 'social' and 'human' concerns. The second largest group (8 out of the 29) is made up of 'monetized' synthetic indicators with a predominantly environmental slant. It would appear, therefore, that those concerned primarily with defending the social dimension of economic well-being and development have not sought to monetize the variables on which their indicators are based, whereas the converse applies to the 'ecologists'.

New indicators and national accounts

These new approaches are currently regarded with circumspection – and that's a euphemism – by economists and specialists in national accounting. It is true that they do not provide credible alternative solutions (this is not their aim) and that national accountants are quite right not to want to throw the baby out with the bathwater. The problem should be framed differently. The aim of these new approaches is not to produce alternatives that can be set up in opposition to national accounts but rather to put the current method of national accounting into perspective and to incorporate it into multidimensional problematics. This does not of course exclude attempts to extend national accounting systems in reasonable ways or even to improve them from within. Most of the innovative indicators that have been developed since 1990 are based to a large extent on national accounting data. Their originality lies in the fact they are not based *solely* on such data.

II
GDP and Growth Called into Question

This chapter examines the notions of GDP and growth in a way that is deliberately simplified and accessible to someone with absolutely no prior knowledge of national accounting or economic statistics. The calculation of economic growth is based on the definition of what is known as the gross domestic product, or GDP. GDP is made up of two parts. The first is the market value of all the goods and services sold in a given country during a year (to be more precise, we should say the market value *added*, but we can simplify this, since it does not affect what follows in any way). This market value is then supplemented by a second part, which is the cost of producing the non-market services provided by government bodies: state education, central and local government services, and so on. Thus – and this is a crucial point – the creation of economic wealth measured in this way, that is using GDP, is a flow of purely market and monetary wealth. And growth is defined as the increase in GDP, that is the increase in the volume of all goods and services that are sold or costed in monetary terms and *produced by paid work*. Once again, this is a simplification, since in order to evaluate GDP in volume (or in 'real') terms, price variations have to be neutralized. Once again, however, this has no effect on what follows.

No further knowledge is required in order to understand what the main issues at stake are. This way of measuring national wealth has three major consequences, which will be illustrated in what follows by means of examples:

Box II.1   Production as defined for the purposes of national accounting

According to experts in national accounting, their aim in measuring GDP is to evaluate in the best possible way the productive contribution of economic activities. Consequently, the contribution of each industry or area of economic activity is measured on the basis of the value of the goods and services produced (in market activities, this means the sales of these goods and services) minus the value of the intermediate goods and services destroyed or transformed during the production process (what is known as intermediate consumption). This gives the gross value added (GVA) for each industry or area of activity, the sum of which for all sectors gives the GDP for the country as a whole.

Beside intermediate consumption, each industry also makes use of fixed capital (machinery, buildings, etc.). In order to take account of this, accountants deduct from the GVA the depreciation these fixed assets undergo as a result of being used. This gives the net value added, the sum of which for all sectors gives the Net Domestic Product (NDP) for the economy as a whole.

Studies that seek to measure wealth as accurately as possible, particularly when their aim is to take account of the consequences of productive activity (especially for the environment), attempt to evaluate the depreciation of our 'endowment in environmental capital'. In this sense, they are attempting to 'supplement' or 'compete with' NDP rather than GDP, as Nordhaus and Tobin did in their time. Nevertheless, it is GDP that is the focus of media attention and which today serves as a quasi-universal reference point for most analyses.

- Everything that can be sold and has monetary value added will bump up GDP and growth, irrespective of whether or not it adds to individual and collective well-being.
- Many activities and resources that contribute to well-being are not counted, simply because they are not market activities or resources or because they do not have a direct production cost expressed in money terms.

- GDP measures only outputs, that is the quantities produced. It takes no account of outcomes (the results in terms of the satisfaction and well-being produced by the consumption of these goods), which are more important in evaluating progress. It indicates how much 'having' and 'producing' there is in a society, not its well-being. Nor does it take account of the distribution of the wealth it measures, or of inequalities, poverty, economic security, and so on, despite the fact that they are virtually unanimously regarded as aspects of a society's well-being.

1 The arguments advanced by national accountants

Faced with these criticisms, the vast majority of national accountants adopt a simple and, in many respect, legitimate line of defence: we know perfectly well that GDP and growth do not measure well-being. 'That's not what they're intended to do. Your accusation is unfounded! Nobody's stopping you suggesting indicators of well-being in addition to GDP and using them! Just leave us to get on with our work and you do whatever you want to do, but get out of our way! We might just possibly consider using "satellite accounts" (see Box II.2), but they are simply not suitable for inclusion in the "central framework".'

These economists and accountants are both right and wrong. They are right, because the current measure of GDP is indeed not designed to assess well-being. It is not an indicator of well-being (although a number of the methods used and changes it has undergone over the past fifty years demonstrably do not wholly exclude considerations of well-being (Gadrey 2003), or of human development or of sustainable development. Such indicators have, therefore, to be constructed in different ways and in parallel with GDP. However, the accountants are also wrong, on at least three different levels.

Firstly, if the current measure of GDP is not designed for the purpose, what is there to prevent a group that wishes to do so from developing it, from redefining it so that it encompasses the forms of wealth it currently ignores? After all, national accounting has evolved since its early days and some activities or services that were ignored initially have subsequently been included because the

---

Box II.2   Satellite accounts

Those national accountants most receptive to the idea of extending the accounts while retaining the overall coherence of the current framework advocate the development of so-called 'satellite accounts' relating to a range of different areas such as education, health, the environment, social protection, tourism, and so on. Such accounts already exist, but their impact in politics and the media seems to have been negligible to date. It is to be feared that these satellite accounts will, like other satellites before them, end up by self-destructing in the public space. However, one may also nourish the hope that they will gradually win favour. In our view, these accounts are to be encouraged, particularly if they can offer a different perspective to that obtained by revolving indefinitely around the 'planet' of the central accounts. They could constitute an intermediate level, conceived with a view to laying the groundwork for future integrated national accounts of sustainable human development (see our conclusion, page 117). This would mean that the lengthy, painstaking job of producing satellite accounts could be incorporated coherently into an alternative vision of wealth, since it would be less narrowly focused on each 'satellite body' of economic and social life, whereas 'health', for example, requires a more holistic approach.

'wealth conventions' have changed (Gadrey 2003). Why can other enlargements not be considered?

Secondly, why do economists and national accountants (encouraged by politicians, who commission these statistics) devote a major part of their efforts to measuring and using GDP and economic growth instead of expending, for example, only 50 per cent of their efforts in that direction and dedicating the other 50 per cent to the development and diffusion of alternative indicators that better reflect notions such as well-being, human development, social development, sustainable development and so on? Are they not imprisoned in a very narrow vision of what really counts and what should be counted?

Thirdly, economists and national accountants state that they do not confuse economic growth with growth in well-being and that if others confuse the two, it is nothing to do with them. That may be true, but if so why is it that, in the media and in public debates, we hear incessantly of economic growth as the indispensable linchpin of progress, while there is little if any mention of indicators of social health, of well-being or of environmental protection? Do economists have no responsibility at all for this unbalanced situation in which a country's health is judged virtually exclusively in terms of its economic health? A headline in *Le Monde* on 18 November 2003 ran as follows: 'Japan on the up, but the Japanese still poorly'. How have we reached a situation in which a country's health can be judged to have improved while the situation of the people who live there is deteriorating? Is this not due to excessive concern with just one of the dimensions of national wealth?

2 GDP, growth and well-being

Let us turn now to the main arguments and examples that will help us better understand how far removed the notions of GDP and economic growth are from those of well-being and development.

The damage caused by the current growth model is not deducted

By way of a first example, let us take a society in which there are many road accidents, which require medical care, vehicle repairs, emergency services and so on. Such a society will tend to have a higher GDP than a society in which people drive carefully (all other things being equal, of course). More precisely, it will tend to direct a large share of its economic resources and activities towards the repair of damage, *without any overall increase in well-being*, rather than towards the production of additional well-being.

Imagine this situation: if a country paid 10 per cent of its population to destroy goods, make holes in roads, damage vehicles, and so on, and another 10 per cent to make good the damage, it would have the same GDP as a country in which the same 20 per cent of employment (whose effects on well-being cancel each other out) was given over to improving health, life expectancy, educational levels and participation in cultural and leisure activities.

The same idea can be applied to expenditure incurred in making good environmental damage linked to human activity. This is what ecologists call defensive expenditure. From this perspective, expenditure (and the corresponding output) incurred in repairing the damage caused by human actions should not be counted as a positive contribution to 'real' wealth. If such damage (pollution, crime, road accidents) reduces well-being and makes it necessary to produce goods and services to the value X in order to repair the damage or provide protection against the risk of damage, there can be no question of X being counted as a positive item in any measurement of 'real' wealth. And since the conventional measure of GDP counts the compensatory output X as a positive item, which is acceptable from a purely economic perspective, X must be deducted from GDP in order better to identify 'real' wealth (that which contributes to well-being). If households are purchasing more and more anti-burglary equipment or anti-pollution devices in response to increasing risks, the corresponding expenditures should be deducted from GDP (or their standard of living) if it is desired to depict the variations in their well-being more accurately. There is no need for such equipment in countries in which the incidence of burglary or of pollution is very low. As Fred Hirsch (1976) wrote in a superb book, if the outside temperature drops and the heating is turned up in order to maintain the inside temperature at a constant level, there is no increase in well-being. This holds true if 'outside temperature' is replaced by 'pollution, crime, accidents, uncontrolled urbanisation, etc.' and 'turning the heating up' by 'increasing reparative or defensive output'.

Let us take a second example. The organized destruction of the Amazon rainforest is an activity that increases global GDP. Nowhere is any account taken of the resultant loss of natural resources, or of the various effects on climate, biodiversity, the long term and the needs of future generations. GDP takes no account of losses of natural resources. On the contrary, it counts the organized destruction of such resources as a positive item on the balance sheet. Similarly, a company that pollutes a river in order to ensure its own economic growth and thereby contribute to GDP causes damage that reduces some people's well-being. Such damage is not included in any measure of economic wealth.

Box II.3   Externalities

For economists, production and consumption activities generate positive or negative 'externalities'. These externalities are negative when this production (or consumption) activity has an incidental or unintended effect (one that is not part of the production or consumption objectives) that harms third parties, but its cost is not reflected in the market, with the result that the nuisance is not reflected in accounts. This is the case, for example, with the pollution caused by an industrial activity.

Economists attempt to evaluate the (economic) cost of these externalities by various methods: measuring the observable economic impacts, evaluating the costs of prevention, estimating the parameters at work in price formation (the hedonistic method) and the so-called contingent valuation method (a survey-based method for determining the price respondents would be prepared to pay in order to avoid the effects of an undesired externality).

Within this analytical framework, the state's role is to 'internalize' all or part of the cost of externalities (particularly through the imposition of taxes).

What emerges from all the previous examples is that the damage and destruction (and losses of well-being) linked to economic growth are simply not recorded anywhere. Thus we have indicators that incessantly add and add economic wealth as soon as production and monetary sale take place, without any concern for what is lost on the way, which may not have any market value but may be of enormous value for our current well-being and that of future generations.

Positive contributions essential to well-being are not counted

As well as these examples of the failure to take account of losses of well-being, there are others in which gains, that is positive contributions essential to well-being, are excluded. Here are a few examples.

If people are encouraged or even forced to work harder and harder and to sacrifice leisure and free time in order to achieve high growth rates, this will be reflected only in an increase in GDP, since GDP is not measured in a way that regards an increase in free time as a benefit worthy of being counted. This example has not been selected at random: in the USA, average annual working time per capita has increased by the equivalent of five weeks' work per year since 1980 (204 hours), in contrast to what has happened in virtually all European countries. This is a good example of an essential contribution to well-being, namely free time, that does not feature in national accounts.

Another example of a forgotten contribution is voluntary work, which is not included in the activities that contribute to national wealth as defined for the purposes of measuring GDP because it does not involve any monetary exchange. Do these activities not produce wealth and well-being in just the same way as paid work?

The third and final example of a forgotten contribution is even more important. It is domestic work, which is carried out in the private sphere, mainly by women, which is not the least of the reasons why it is ignored. It is indeed the very epitome of invisible work. Despite this, enormous volumes of such work are done, and there is no doubt that it contributes to well-being in the same way as paid work. It is estimated that the total time spent on unpaid domestic work in developed countries is of the same order of magnitude as that spent on all paid work. If it was decided, for example, to attribute to it the same monetary value per hour worked, GDP could be doubled! And even if a lower value was attached to it, for example that of the hourly cost of a cleaner or a home help, this would still mean that considerable amounts of wealth were being ignored.

GDP is concerned with outputs not outcomes

It is no secret that wealth is not the same as well-being. This latter notion can be approached in two ways. The first is to evaluate subjective well-being by means of opinion polls or satisfaction surveys. These are, it is true, tricky to interpret but they do at least reveal very considerable discrepancies between the evolution of living standards and perceptions of the evolution of well-being. Numerous surveys conducted in several countries have shown that this is so. For example:

when Canadians were asked in 1998 how the overall financial situation of their generation compared to that of their parents at the same stage of life, less than half (44%) thought that there had been an improvement, despite an increase of approximately 60% in real GDP per capita over the previous 25 years.

(Osberg & Sharpe 2003)

The other approach to well-being is to assess objective well-being on the basis of a range of criteria such as good health and life expectancy, access to education and knowledge, economic security, the incidence of poverty and inequalities, housing and working conditions, and so on. Now GDP measures only volumes of outputs (the volume of goods and the quantity of services produced) and is not concerned with these outcomes. For example, the contribution of health services to growth is measured only in terms of the volume of consultations, hospital admissions, treatments and the like and never in terms of the contribution such services make to improvements in health and living conditions. If such a measure were introduced, an effective policy on the prevention of health risks would tend to reduce the contribution of health services to growth, whereas it would probably increase well-being.

Growth for whom?

Besides the question of 'growth for what?', there is also the question of 'growth for whom?', that is the question of inequalities. Depending on the country in question, the same growth rate of 2 or 3 per cent per annum over a number of years may go hand in hand with a widening or a reduction in social inequalities. And yet these phenomena have no part in the prevailing concept of wealth. Is this normal? Does the fact of living in a society in which vast numbers of poor people coexist with a handful of very rich individuals have no impact at all on our well-being? Does not a euro or dollar's worth of growth in a poor man's pocket produce more well-being than the same sum added to Bill Gates' wallet? And yet this is the hypothesis advanced by those who regard GDP, wealth and progress as one and the same thing. And again, while it is true that no national accountant would defend such a comparison, it is clear that it is in widespread use on a daily basis, since the overwhelming dominance of the market and monetary dimensions in evaluations of progress is not counterbalanced by any alternative indicators of similar weight.

III
Human Development and Social Progress

Most of the synthetic indicators we have surveyed are concerned primarily with 'human and social' questions, expressed in terms of human development, 'social health', well-being and quality of life. These terms are not interchangeable. In some cases, the underlying concepts are individualist or even utilitarian, while in others they are much more collective or 'societal', particularly when inequality and protection against economic insecurity are included as criteria. We will not attempt to elucidate the philosophical foundations of each of the initiatives included in our survey. In some cases, the authors explicitly do that job for us. In other cases, the philosophical foundations are implicit or seem to be mixed. We tend to think that the political future of new indicators will depend on their ability to combine individualist concepts with societal objectives, without bringing them into conflict with each other. Incidentally, is this not a requirement of democracy in general?

We have selected four synthetic indicators for examination in this chapter. All of them are concerned with questions of human development and social progress. The best known are those of the United Nations Development Programme (UNDP) and the index of social health.

One point of terminology needs to be clarified. When we refer to the various 'dimensions' of a synthetic indicator, we are talking about the major areas or spheres it encompasses. The 'variables' are the most disaggregated components of each dimension. This applies only to those indicators – the majority of them – that have two levels of aggregation.

25

1 The UNDP development indicators

These indicators, or even the HDI alone, with which the UNDP laid the foundations for the construction of large-scale international indicators and which is still the best known and most widely diffused, are a subject that could be 'developed' almost *ad infinitum*. Indeed, they have already been the subject of many studies and much criticism. In May 1998, the journal *Futuribles* published a clear, well-argued but highly critical article by Jean Baneth, former World Bank economist, which constituted a systematic condemnation of these indicators and their creators. We will mention these criticisms later; indeed, we share some of them, which frequently make good statistical sense. However, we have absolutely no sympathy at all with the general thrust of this attempt to demolish an important collective enterprise that has shown a capacity for self-questioning and improvement.

As for the charge that a little more weight has been accorded to 'third-worldist' ideas in perfecting the development indicators, this is nothing more than a modest and wholly justifiable attempt to shift the balance of power somewhat.

Four synthetic indicators

Since 1990, the UNDP has published a *Human Development Report*, which contains a battery of economic, social and environmental indicators that have been extended and refined over the years. The Report includes the celebrated, albeit rudimentary Human Development Index, whose diffusion throughout the world has been a spectacular success beyond the developing countries for which it was originally intended. This indicator is quite simply the average of three indicators that classify countries on a scale of 0 to 1: GDP per capita (expressed in purchasing power parities or PPP); life expectancy at birth; and a knowledge index (measured by a $\frac{2}{3} - \frac{1}{3}$ combination of the adult literacy rate and the gross enrolment ratio). Box III.1 clarifies certain points of methodology concerning the calculation of the HDI.

The UNDP has subsequently published three other synthetic indicators on an annual basis. The first of these, the Gender-Related Development Index (GDI), which provides a basis for evaluating differences in the situation of men and women in terms of the three criteria used to characterize human development, was introduced in 1995. The same year also saw the introduction of the Gender Empowerment Measure (GEM), which supplements the GDI. The third indicator, the Human Poverty Index (HPI), was first published in 1997. It records the basic shortages, privations and exclusions suffered by a section of the population, with variant 1 applying to developing countries and variant 2 to developed countries. For developed countries, the HPI-2 is based on four criteria, all given the same weighting: probability at birth of not surviving to age 60, percentage of adults lacking functional literary skills, percentage of people living below the poverty line and long-term unemployment rate.

Box III.1   Calculating the HDI

The HDI is itself based on three sub-indexes, each with a value between 0 and 1, which are averaged to produce the global index. However, the second sub-index is itself a weighted average of two component indexes. The three sub-indexes are described below. In each case, the method for calculating a value between 0 and 1 is outlined.

1. **Index of life expectancy at birth:** Life expectancy at birth (E) is expressed in years. In order to convert it into an index (Iexp), the following formula is used: $Iexp = (E - 25)/(85 - 25)$. 85 is a maximum value that no country currently reaches and will probably not reach for a long time (although the figure for Japan is already over 80) and 25 is a value significantly lower than those of the (African) countries where mortality is highest (their life expectancy is around 35 years). Example: in France in 2001, life expectancy at birth was 78.7 years, which gives an index of 0.895.

2. **Education index (literacy + enrolment):** The education index (Ied) measures the level attained by a country as reflected in the adult literacy rate and enrolment rates in primary, secondary and tertiary education (gross combined enrolment rate). The procedure involves, firstly, calculating an index for adult literacy (share of the adult population that

(continued)

Box III.1   *continued*

is literate, 0 to 1) and another for enrolment (numbers enrolled in the three educational levels divided by the total population of corresponding age). These two indexes are then merged (weighted average) to give the education index, in which adult literacy is given a $\frac{2}{3}$ weighting and the gross combined enrolment rate a $\frac{1}{3}$ weighting.

This method makes sense in countries where literacy is a significant problem and is regularly monitored. It makes much less sense in more developed countries, where reliable statistics are often not available anyway (as is the case in France). In such cases, the UNDP imputes a standard value of 0.99. The education index for France in 2001 was 0.964.

3. **Index of GDP per capita:** The GDP index (Igdp) is calculated on the basis of per capita GDP (in purchasing power parities)[1] 'adjusted' by a log function (base 10). This adjustment is based on the following notion: unlimited income is not necessary in order to reach an acceptable level of human development. By introducing a log function, therefore, those responsible for the index are adopting the hypothesis that economic growth has diminishing returns in terms of human development, all other things being equal as far as the two other dimensions are concerned. This hypothesis is further reinforced by another convention on the upper and lower thresholds required to convert the adjusted value for GDP into an index between 0 and 1 (these thresholds play the same roles here as the 25- and 85-year thresholds for the Iexp). The lower threshold for per capita GDP is set at $100 per year and the upper threshold at $40,000. Box III.2 (page 30) presents a strong criticism of such conventions.

For France, whose per capita GDP was 23,990 (PPP) in 2001, the GDP index is 0.91. It is calculated as follows: GDP index = [log (23 990) – log (100)] / [log (40 000) – log (100)] = 0.915. The HDI is the simple average of these three sub-indexes (Iexp + Ini + Igdp)/3. It was 0.925 for France in 2001.

Table 1 below shows the ranking of the 20 leading countries (only 17 in the case of the HPI, for lack of data for some countries such as Switzerland and Austria, which are, nevertheless, wealthy) by, respectively, the HDI, per capita GDP, the HPE and the GEM as they appeared in the 2004 UNDP report. France does not figure in the GEM classification, since it has not authorized publication of these figures. In a sense, it is better this way, since France would be given a very low ranking, which is probably why it refuses to authorize publication.

Table 1   Developed countries ranked by four indicators

HDI ranking (ranking by per capita GDP [PPP] 2002)	Poverty: HPI-2 (2002)	Gender empowerment GEM (2002)
1. Norway (2)	1. Sweden	1. Norway
2. Sweden (20)	2. Norway	2. Sweden
3. Australia (11)	3. Netherlands	3. Denmark
4. Canada (8)	4. Finland	4. Finland
5. Netherlands (10)	5. Denmark	5. Netherlands
6. Belgium (12)	6. Germany	6. Iceland
7. Iceland (7)	7. Luxembourg	7. Belgium
8. USA (4)	8. France	8. Australia
9. Japan (14)	9. Spain	9. Germany
10. Ireland (3)	10. Japan	10. Canada
11. Switzerland (6)	11. Italy	11. New Zealand
12. UK (19)	12. Canada	12. Switzerland
13. Finland (18)	13. Belgium	13. Austria
14. Austria (9)	14. Australia	14. USA
15. Luxembourg (1)	15. UK	15. Spain
16. France (15)	16. Ireland	16. Ireland
17. Denmark (5)	17. USA	17. Bahamas
18. New Zealand (22)		18. UK
19. Germany (13)		19. Costa Rica
20. Spain (23)		20. Singapore
21. Italy (17)		21. Argentina

Notes: Reading the table: for each of these four indicators, the country ranked highest is the one that performs 'the best'. Thus in terms of poverty, Sweden is ranked first in the sense that it is the country where, according to this indicator, there is least poverty.

Source: UNDP Report 2004

Box III.2   Indicators ill-suited to developed countries?

Table 1 does not show the values of these indicators for the countries in question. However, a few observations would be appropriate here. After all, there is a difficulty, and one that tends to support some of the criticisms made by Jean Baneth and others. The HDI and the GDI are unable to show significant differences between *developed countries*, although there are good reasons to believe that such differences exist. There are various explanations for this relative inability, but they all have to do essentially with the method outlined in Box III.1 by which a score of between 0 and 1 is calculated for each component of the overall index (per capita GDP, life expectancy and education). This is particularly clear in the case of per capita GDP, where a debatable convention has been adopted. A score of 1 is imputed when a country reaches the level of $40,000 (in terms of PPP). Furthermore, as countries approach this threshold, which is the case with the wealthiest countries, the increase in per capita GDP has hardly any influence any longer on the HDI. For example, when per capita GDP rises from $25,000 to $30,000 (most of the 15 wealthiest countries have reached such levels), the component of the HDI related to this variable increases from 0.922 to 0.951 and it rises from 0.951 to 1 (the absolute maximum) when per capita GDP increases from $30,000 to $40,000. Currently, only one country, Luxembourg, has exceeded the $40,000 level.

There is a fairly sound notion underlying this convention, namely that economic wealth has 'decreasing returns' for human development or well-being. Is it necessary to go so far as to transform this sound idea into an absolute 'ceiling' above which growth makes zero contribution to human development? This is not evidently the case, and the same solution could be adopted for the HDI as the UNDP adopted for poverty, which was to construct two poverty indexes, one for developed countries and one for developing countries. This would reintroduce some differences between developed countries, while at the same time retaining a good proportion of the sound ideas that make the HDI and GDI well suited to their primary task, which is to analyse the performance of countries with low or average levels of human development, that is the vast majority of countries in the world (128 of the 174 countries included in the UNDP statistics).

(continued)

Box III.2 *continued*

As far as the human poverty index (the only one measured in terms of shares of the population rather than a scale of 0 to 1) and the gender empowerment measure are concerned, they do not suffer from the disadvantages outlined above. They are much more discriminative within the group of developed countries, undoubtedly because, in these areas and to some extent irrespective of economic wealth per habitant, many 'advanced countries' are lagging a way behind the 'best' and are not making much of an effort to catch up.

Whatever the limitations of these indicators might be, they certainly 'indicate' many things, even for developed countries. It is not unimportant, for example, to observe that the Nordic countries achieve excellent scores in virtually all the categories, particularly with regard to the reduction of various forms of inequality (such as poverty, inequalities between men and women), while at the same time obtaining very respectable rankings when economic wealth (HDI) comes into play (to some extent). Nor is it without interest to note that there are countries whose social performance (in terms of rankings) is significantly better than their raw economic performance (the Nordic countries again) or that the four countries ranked lowest by the poverty criterion (albeit in a list limited to only 17 countries) are, in descending order, Australia, the UK, Ireland and the USA, which are all countries characterized, in varying degrees and in their different ways, by the so-called 'Anglo-Saxon' social model and its values.

Finally, if we wish to dig a little deeper, we cannot confine ourselves to these large-scale synthetic indicators. They are an encouragement to track down, within the UNDP tables, their numerous constituent variables, from which much can be gleaned. At the same time, however, any comparative approach based on statistics that seeks to encompass all the countries in the world will come up against problems of relevance and reliability, particularly when it comes to classifying the developed countries. Consequently, we need to examine other approaches which, by virtue of being less than global in ambition, are able to go further in their evaluations of each country or a limited number of countries. We will come to them in what follows, without seeking to pit the two approaches against each other. They are two complementary ways of looking at the world.

Three synthetic 'social' indicators will be outlined in the following sections. The characteristic they all share is that they are concerned primarily with questions of inequality, poverty and various 'social pathologies'. They are summaries of 'major social problems', as seen by those responsible for researching and designing the indexes.

2 The index of social health

This index was developed at the Fordham Institute for Innovation in Social Policy (Fordham University, Tarrytown, NY) by Marc and Marque-Luisa Miringoff, whose first studies in this area date from the second half of the 1980s. Their index of social health (ISH) has become very well known internationally since 1996, the year a major article appeared in *Challenge*. Its reputation was further enhanced by the subsequent publication of a book in 1999. The ISH has also been applied, with a few modifications, in Canada (Brink & Zeesman 1997) and in the State of Connecticut.

The ISH is a synthetic social indicator intended as an alternative or supplement to GDP in making evaluations of progress. It is calculated on the basis of 16 basic variables allocated to one of five groups associated with particular age categories. Table 2 demonstrates the construction of the index. The value of an approach based on age categories is explained in the following terms by Brink and Zeesman (1997, p. 11):

- Age groups are universal, with everyone potentially passing through all age groups.
- Age groups are conceptually integrated across components, creating an holistic framework.
- Age groups highlight several important contemporary social trends, such as the deteriorating status of children and the improving status of the elderly during the decade of the eighties.
- Age groups are readily understood by the public, facilitating policy discussions which might result from the Index.

Table 2 The components of the index of social health

Children	Youth	Adults	Elderly	All ages
Infant mortality	Youth suicides	Unemployment	Persons 65 and over in poverty	Violent crimes
Child abuse	Drug abuse	Average weekly earnings		Alcohol-related traffic fatalities
Child poverty	High school drop-outs	Health care coverage	Life expectancy, aged 65	Access to affordable housing
	Teenage births			Gap between rich and poor

Box III.3 The aggregation method used for the ISH

It is useful at this point to examine in some detail the aggregation method, since it raises some awkward questions, both for the ISH and for the other indicators outlined in this chapter. In this method, each variable is given a score for each year; the annual scores are then averaged out. The procedure is as follows. For each variable, a score of 0 represents the worst performance in the period under investigation, while a score of 100 represents the best performance. In other words, and this is an essential point, there is no attempt to define norms for the 'best' and 'worst' performance that can be achieved (for example by seeking out such norms in the countries that perform the best and those that perform the worst, or by adopting an even more normative approach). The intermediate results are obtained by means of a 'linear interpolation'. This method (adopted in France by the initiators of the BIP 40) is a practical one and is not devoid of sense, but it does have various disadvantages, the following three in particular. Firstly, if performances for one variable remain poor throughout the entire period (for example, an unemployment rate fluctuating between 10 and 15 per cent), the score of 100 will be awarded for a bad performance (an unemployment rate of 10 per cent). This is not really a problem as long as the index is used only to monitor variations over time. Secondly, if the index is subsequently recalculated over a longer period (for example, by substituting 1970–2000 for 1970–90), the 'bases' for attributing the scores of 0 and 100 are

(continued)

Box III.3   *continued*

very likely to change for some of the variables. Again, this is not a dramatic problem. Even national accountants have to 'change base' periodically and recalculate everything. The only major problem is the third one, and the difficulty is twofold.

1) If a variable changes only slightly over the period, a slight increase in, say, life expectancy (imaginary example: over 30 years, life expectancy increases regularly and rises from 65 years to 65 years and 1 month) will bring the score for this variable from 0 to 100. This will have as much weight in the final index as if unemployment, for example, had fallen from 12 per cent to 5 per cent, which everyone would regard as much more significant in terms of social progress.

2) If all the variables increased (or declined) by 1 per cent, this would produce the same curve as if they increased (or declined) by 10 per cent, or any other amount.

Attempts can be made to get over these difficulties by various fairly arbitrary techniques. However, it is not unreasonable – not least to avoid muddying the waters further – to stick with an indicator that is very imperfect but is better suited than others to the purpose of making an initial judgement of social progress.

The *Challenge* article (Miringoff & Miringoff 1996) was the first to present variations in GDP and those in the new index on the same graph. The divergence of the two indicators from about 1973 onwards (with GDP continuing to increase and the ISH declining significantly and on a sustained basis) produced a spectacular visual effect. This graph is reproduced here (Figure 1, page 37) in a version, based on just nine indicators and updated to 1996, which is the only one to go as far back as 1959. There is a need for caution in interpreting such graphs, however (see Box III.4 pages 36–38).

In the case of the ISH, as in that of the UNDP indicators or the BIP 40 (see below), no attempt to produce a relevant diagnosis of the evolution of 'social health' can sensibly confine itself to the synthetic index: the component indicators and their variations also need to be scrutinized. The synthetic index, which constitutes a fairly crude summary of the available data, is all the more useful to academic and public debate if it is regarded simply as one stage in the evaluation process, an encouragement to spend far more time breaking down the problems it flags up.

The largest part of Marc and Marque-Luisa Miringoff's excellent book (1999) is given over not to the synthetic index of social health but rather to analysis of the component indicators and of other, closely related indicators, which are allocated to one of three categories: those that advanced between 1970 and 1996 (4 of the 16), those that declined (7) and those whose variations fluctuated without any clear dominant trend (5). In short, and this is true of all the other synthetic indicators as well, the most spectacular results the ISH produces – and the ones most likely to be seized on by the media, particularly the comparison with GDP – are certainly the ones that are most questionable academically. Nevertheless, they have the enormous merit of attracting attention to questions which, for lack of such attempts, might well never 'hit the headlines', although they are just as important, if not more so, than the health of the economy and stock exchange prices.

In view of the possible application of the ISH to other countries, there is one final remark to be made, particularly since efforts in this direction seem already to be under way. Obviously, the ISH is, in part at least, what Brink and Zeesman (1997) describe as 'country-specific'. The 'social pathologies' that are recognized as such and measured depend on the institutional and cultural context. The 'major social problems' are hierarchized differently from country to country. True, each of the 16 variables shown in Table 2 would be meaningful in other countries, but it is highly likely, given the need to focus on a reasonable number of criteria, that national debates on the variables regarded as the most important would produce lists that overlapped to some extent but also differed in various respects as well. And since the objective here is not to conduct an international comparison of the values of the synthetic index but rather to monitor its evolution over time, different variables would have to be selected if an ISH were to be constructed for a different country.

Box III.4   The need for caution in interpreting graphs such as the ISH

The ISH provides us with an opportunity to outline some initial observations on this type of comparison, often illustrated by eye-catching graphs, between the traditional economic indicators derived from national accounts (GDP, consumption etc.) and various synthetic indicators that are assumed to reflect more accurately the evolution of 'social health', well-being or quality of life. After all, there is a risk that the visual impressions produced among individuals with little grasp of the possible significance of such variations could be used for demagogic purposes. In particular, these graphs maintain the illusion that since, under 'normal' circumstances, the economy and social health should grow in tandem – a norm that certainly meets with fairly broad unanimity, whether sincere or hypocritical – the two curves should evolve in parallel with each other. It is obviously a very bad sign if the first advances but the second stagnates or declines. However, it is very difficult to express an opinion about the situations, of which there are many after all, in which the index of social health progresses, but more slowly than economic growth (or than per capita GDP or consumption).

The reasons for this are very simple. On the one hand, economists are of the view that economic growth can in theory continue at high rates for centuries, although this is in fact open to doubt. On the other hand, virtually all the social indicators that make up the ISH are 'limited' by their construction. It is impossible for unemployment, suicide and poverty rates to go below 0 per cent and the Gini coefficient, which measures income inequalities, can never go below a value of 0, which equates to a perfectly equal income distribution. Under these conditions, social progress *measured in this way* will, over the long term, necessarily come up against a 'horizontal asymptote', which is not the case with economic growth, at least not within a foreseeable time horizon. The researchers at the Fordham Institute have certainly tried to get round this difficulty by adopting the method, outlined above, of attributing a score of between 0 and 100 to each *period under analysis*, which makes it possible to reintroduce the relative differences when the absolute differences decrease. However, such an exercise obviously has serious limitations and does not really resolve the issue. It is, and always will be, difficult to interpret the differing paths taken over long periods by synthetic curves

(continued)

Figure 1 Index of social health (nine indicators) and GDP (base 50 in 1959), in 1996 prices: USA, 1959–96 (after Miringoff & Miringoff 1996)

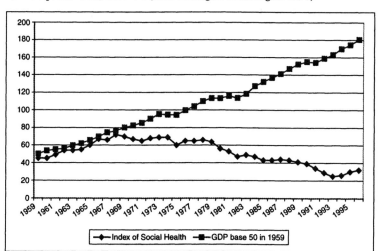

Box III.4 *continued*

representing, independently of each other, 'unlimited' economic performances and social performances that are given scores of between 0 and 100. Readers examining Figure 1 should be aware that even if all the various aspects of social health have progressed considerably in the United States over this period, the ISH curve would have tended towards the 100 mark without ever exceeding it and there would in any event have been a certain degree of divergence or 'decoupling' from the GDP curve.

Does this warning lead to an impasse? Not necessarily.

Firstly, the social indicators in question, like the traditional economic indicators, are historical constructs whose validity is restricted to certain periods of history and to certain visions of social progress and its priorities. Under these conditions, it is of little significance that an indicator like the ISH may well, in the long term, 'come up against an asymptote'. At the moment, we are sufficiently far away from the asymptote for the problem not really to have arisen yet, and this situation may very well persist. This indicator will not remain pertinent for ever, but then nor will the others. It is sufficient for it to be considered relevant (on the basis of social conventions) for the period under investigation. The problem of the 'asymptotic stagnation' of such and such an indicator

(continued)

Box III.4   *continued*

merely reflects the fact that we are dealing here with concepts that are all historically dated, including GDP, and have 'zero asymptotic relevance'. This should in no way inhibit their use during the period of their 'social life cycle', when they are relatively significant.

These arguments could be supplemented by a slightly more debatable but nevertheless defendable one. It may be true that economic growth can continue for decades at a rate of 2 per cent or more and that the social indicators most commonly referred to have absolutely no chance of being able to do so, which seems to put them at a disadvantage statistically. However, there are also grounds for believing that the returns of economic growth (which is 'quantitative' by virtue of its very construction) in terms of well-being and human development start to decline once living standards and overall wealth have reached certain thresholds, despite persistent inequalities of distribution, which mean that these 'returns' are not the same for everyone. There are even those who believe that, beyond certain limits, surplus growth would lead to a decline in the quality of life or of 'sustainable' well-being. Thus if the objective of most of the synthetic socio-economic indicators is to provide the tools for a more sophisticated approach to the evolution of well-being and human development (or similar notions), 'raw' economic performance (increase in production or consumption) should not be taken into account directly but should be replaced by a component adjusted for the contribution it makes to well-being. In other words, the fact that the standard social indicators seem intrinsically narrow or limited (because of the way they are defined, which is itself very quantitative), and that they are therefore *disadvantaged* in the statistical competition, simply reflects the fact that 'raw' economic performance is unduly *advantaged* when the 'economic' and 'social' curves are merely juxtaposed. These observations could well be seen as an argument in favour of socio-economic approaches that combine a 'corrected' measure of GDP (or of consumption), along the lines of the indexes of 'sustainable well-being' (see Chapter V), and a synthetic social indicator such as the ISH, with care obviously being taken to avoid counting the same variables twice. In this respect, the studies by Osberg and Sharpe (see Chapter VI) constitute an interesting compromise.

3 The BIP 40: a barometer of inequalities and poverty in France

The notion that the health of the economy and of the stock exchange needs to be reflected in widely publicized synthetic indicators, while 'social health' gets the smallest share of media attention, was also the mainspring of efforts by certain economists and statisticians in France associated with a network that campaigns for a reduction in inequality to develop and present to the press a new synthetic indicator, the BIP 40, in 2002.

The aim of this indicator is to cover several dimensions of inequality and poverty, to construct an indicator (itself the result of several other indicators) for each dimension that can be used to track the evolution of the corresponding inequality over time and, finally, to sum (or 'aggregate') these various indicators in order to produce a global indicator (the BIP 40). Box III.5 explains the method.

Let us start with the selected dimensions and their content. They total six in all:

1. **Employment and work**: the 24 indicators used to evaluate this dimension are divided into four groups: unemployment (eight indicators, including the overall unemployment rate, as well as gender differences in unemployment rate and the share of long-term unemployment), precarity (five indicators), working conditions (eight indicators) and industrial relations (three indicators).
2. **Earnings**: there are 15 indicators for this dimension. Again, they are divided into four categories: wages (inequalities, share of low-paid workers, etc; five indicators in all), poverty (four indicators), inequalities and the tax system (three indicators) and consumption (three indicators).
3. **Health**: a total of five indicators, similar to those used by the UNDP in its annual human development reports (e.g. life expectancy, differences in life expectancy between various occupational categories, etc.).
4. **Education**: five indicators in all, including the share of young people leaving the school system without qualifications and some measures of inequality in educational attainment.
5. **Housing**: five indicators, including the share of social or subsidized housing in new builds.
6. **Justice**: four indicators, including the incarceration rate.

Box III.5   The method used to aggregate the BIP 40

What is the procedure for summing the indicators used within each dimension and then the indicators for each dimension in order to obtain a single overall figure? The answer is that it is the same as that used for the ISH, except for the final stage, when the weightings are introduced (the ISH just uses the simple average). Let's take a closer look. The procedure consists of two separate operations. The first involves reducing all the partial indicators to a single 'score' of between 0 and 10. To this end, a score of 0 is awarded for the best performance during the period under investigation (in this case between 1982 and 2002) and a score of 10 for the worst. This choice means that the index rises when inequalities and poverty increase. For example, if the lowest unemployment rate during the period in question was 8 per cent and the highest was 12.5 per cent (which was more or less the case in France between 1982 and 2002), a score of 0 would be attributed to the value of 8 per cent and a score of 10 to the 12.5 per cent figure. The intermediate values are scored by linear interpolation (rule of three). Thus a figure of 10 per cent (recorded for 1992 and 2000) will be scored thus: $(10 - 8)/(12.5 - 8) \times 10 = 4.4$.

   The second operation involves 'aggregating' these indicators, all of which have now been transformed into scores of between 0 and 10, firstly within each of the six dimensions and then for all of the dimensions. It would be possible on each occasion simply to take the average. However, there are good reasons for thinking that some indicators are more important than others in the public debate on inequality and poverty. It is assumed, therefore, that greater 'weight' has to be given to some indicators than to others, which leads to the calculation of a 'weighted average'. Obviously, the choice of the weighting coefficients is to some extent arbitrary (even if the same weighting were to be chosen for each dimension),

(continued)

Box III.5 *continued*

but the arbitrariness can be reduced by discussing the choice in a way that takes account of society's preferences. For example, the group that developed the BIP 40 decided to give the first two dimensions (work and employment and earnings) twice as much weight as the others. This seems reason able, since inequalities in earnings, employment and work obviously play a decisive role in any assessment of social inequalities. Figure 2 shows the evolution of the overall index between 1982 and 2002. The BIP 40 rises rapidly from 1983 onwards. It marks time between 1990 and 1993 before taking a definite upward turn until 1998. Between 1998 and 2001 things improved, with poverty and inequality both declining. However, they began to rise in 2002 and, in all probability, into 2003. It remains the case that, according to this index, inequality in France was considerably greater in 2002 than at the beginning of the 1980s. The data for each of the six dimensions, which are not reproduced here, show that the deterioration was particularly pronounced in the 'work and employment' dimension, that the synthetic indexes for earnings, health and education remained fairly stable but that inequality increased in the housing and justice dimensions. Here too, however, it is useful to examine in detail the variables that make up each of the six dimensions. For example, while it is true that the synthetic health indicator did not deteriorate overall, this is largely because the constant increase in life expectancy, which has the effect of raising the overall health indicator, is offset by an inequality variable whose evolution reflects a spectacular social decline: the difference in life expectancy between managerial employees and manual workers, which was 'only' 4.8 years in 1982, had risen to eight years in 2002!

Figure 2   The BIP 40, 1984–2002 (*Source*: www.bip40.org)

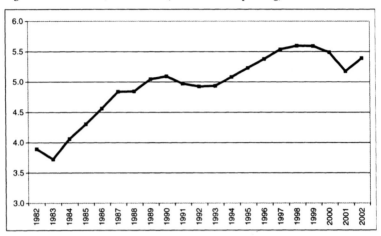

We should note here the birth, in 2003, of a Belgian cousin of the BIP 40, the Index of Social Insecurity, developed by the Institute for Sustainable Development (see the list of websites at the end of the book).

4 The Personal Security Index (PSI)

Developed in the mid-1990s by the Canadian Council on Social Development (CCSD), the PSI has the advantage of incorporating aspects that are not included in the indicators examined to date. Security is regarded as a key element in the perception and measurement of well-being. The conception of security adopted here is multidimensional and encompasses three major dimensions:

(a) economic security (a dimension also adopted by Osberg and Sharpe), which includes employment security, safety net and financial vulnerability;
(b) health security (protection against the risks of illness); and
(c) physical security (feeling of safety with regard to crime).

In conceptual terms, the aim is to evaluate individuals' quality of life more accurately in terms of the various forms of insecurity to which they are exposed by constructing a single indicator designed to make a substantive contribution to public debate.

From a methodological point of view, the synthetic indicator aggregates heterogeneous data compiled on the basis of a method that combines the principle adopted for the ISH and that used for the Kids Count index (see box). However, its originality lies mainly in the fact that it is one of the few indicators that combines objective and subjective aspects of well-being. Thus by publishing the data and tracking their evolution (the first indicators were published in 1998), not only can this indicator be compared with the overall trends in economic growth but the differences between the 'objective' data and individuals' perception of the various forms of insecurity can also be examined. The indicator can also be used to compare the six Canadian regions by sex and by age group. Box III.6 explains the method used to construct the PSI. (On objective and subjective indicators, see also Box I.2, entitled 'Other criteria for differentiating the indicators'.)

This methodological innovation comes at a price, however, since it requires that an opinion survey be carried out each year. Although the CCSD is an independent institution, the survey is funded by two Canadian government departments, Canadian Heritage, which is responsible for multiculturalism and social cohesion, and Health Canada.

Table 3   PSI: the variables and their weighting

Data index	Weighting	Perception Index	Weighting
Economic sub-index	*35*	*Economic sub-index*	*35*
Disposable income	5.83	How adequate would you say your household income is in meeting your family's basic needs?	8.75
Average poverty gap	5.83	I think there is a good chance I could lose my job over the next couple of years	8.75
Long-term unemployment (over 12 months)	5.83	If I lost my job, I am confident that I could find an equivalent one within six months	8.75

Table 3   PSI: the variables and their weighting – *continued*

Data index	Weighting	Perception Index	Weighting
Employment insurance coverage	5.83	If I lost my job, I am confident that I could count on government support programs to sustain me and my family adequately while I looked for a new job	8.75
Social assistance level (% LICOs)	5.83		
Level of personal debt (% disposable income)	5.83		
Health sub-index	*55*	*Health sub-index*	*55*
Potential years of life lost	18.33	In general, how would you rate your health?	18.33
Work injuries	18.33	How stressful would you say your life is?	18.33
Traffic injuries	18.33	I am confident that if I or a member of my family were to become seriously ill, we would be able to access the necessary health care system	18.33
Personal safety sub-index	*10*	*Personal safety sub-index*	*10*
Violent crime	5	Thinking of your family's exposure to violent crime, how safe is your neighbourhood?	5
Property crime	5	Thinking of your household's exposure to property crime such as break-ins, how safe is your neighbourhood?	5
PSI data index: grand total	100	PSI perception index: grand total	100

Box III.6   The method used to construct the PSI

In each of the three areas (economy, health and physical safety), there are several indicators based on 'objective' variables and several questions that seek to elicit individual's own perceptions. The weightings used differ from the indicators outlined hitherto in that they are derived from 'subjective data'. In 1999, a large-scale survey was conducted among Canadians in order to ascertain the importance they attached to the three dimensions of their security (economic, health and physical). The responses, which were to be used for a period of at least three years in order to maintain consistency, were as follows: 55 per cent gave top priority to health, 35 per cent to economic security and 10 per cent to physical safety. These were the weightings subsequently adopted for both the objective and subjective data. Within each dimension, however, no specific weighting was allocated, so that all the component variables have the same weight.

As far as the objective aspects are concerned, the standardized changes are those for the years preceding the baseline year (1994–98). The index for 1998, which serves as a reference, was set arbitrarily at 100. Within each dimension, each of the values has been given equal weight; as a result, since the weight of the economic security dimension, for example, is 35 per cent, each of the six constituent lines of this dimension was allocated a value of 5.83 (35/6) for 1998.

As far as the subjective aspects are concerned, all the answers were allocated a value on a scale from 1 to 7 and then averaged. The relative changes in this average between 1998 and the year under analysis are applied to the value for 1998, the arbitrary baseline year.

The annual report, which is 60 pages in length and contains all the objective and subjective data, attaches significantly greater importance to the various components of the synthetic indicator than to the indicator itself and constantly juxtaposes the evolution of the objective data and individual Canadians' perception of each of the dimensions of their security.

5 Local indicators

A book of similar length would be required in order to do justice to the plethora of local indicators that have emerged since the 1990s, mainly in the developed countries, and the community indicators that appeared during the same period in North America.[2] In the vast majority of cases, these are not synthetic indicators but assessments or multi-dimensional 'dashboards' that incorporate a certain number of variables that are virtually the same as those found in the more general initiatives surveyed in this book and which attach very different levels of importance to economic, social and environmental questions. The values that are emphasized are also similar (giving greater priority to social progress, safeguarding the environment, etc.), although in many cases there are questions about the quality of local democracy or governance. The environmental problems that gave rise to so-called local agenda 21s (the term denotes a sustainable development strategy that emerged from the 1992 Rio Conference) are one of the factors that have played a role in the development of this trend.

In some cases, the use of one or more synthetic indicators is part of a political or communication strategy linked to the use of these data. Thus in territories that are of sufficient size for such a project to make sense and for which data are available (a French region, a state in the USA or a Canadian province, for example), some fairly ambitious schemes making use of indicators such as the HDI, ISH and GPI or notions such as the 'ecological footprint' are already under way.

Cities are also involved in these initiatives. Regions and/or cities have also set up various networks in order to exchange experiences (the European REVES network of 'socially responsible territories', for example, the network of sustainable towns and cities in Europe or Canada, or the Aalborg Charter with its strong environmental emphasis, which was drawn up in 1994 and has been signed by a growing number of European towns and cities). The place of indicators in these strategies varies, but they are always present in one way or another.

Of course, these initiatives – like the others – have to be regarded with some degree of circumspection. In particular, we should ask to what extent the craze for indicators is driven by the prevailing fashion for quantification or the familiar reflex that causes the establishment of a statistical unit to be recommended as a means of addressing basic problems that can really only be solved by collective action and the political use of any indicators that might emerge. However, the strength of this trend and examination of its impact suggest that this is indeed a lasting phenomenon, linked both to the localization of public action and other, more general factors associated with opposition to the 'religion' of economic growth and its figures.

We shall limit ourselves here to recommending, in a rather normative way, a 'political method'. The possible benefit to be derived from using local indicators depends on the quality of the local political action of which they form part. Thus, in order to answer the basic question of what constitutes a region's wealth, it is important that the actors work in partnership in order to construct and select the words, values, objectives, criteria, methods of evaluation and judgement and, possibly, the indicators. Another crucial condition is that those advocating the use of local indicators should gradually acquire sufficient legitimacy in the territory in question. Such legitimacy cannot simply be created out of nothing but has to be earned through persuasion, the establishment of alliances and networks of mutual interest, the negotiation of compromises and the intelligent management of conflicting values.

IV
The First Extensions of GDP

The general idea underlying the construction of the indicators that are the subject of this chapter and the next one is that national accounts can be 'corrected' or 'enhanced' by taking into account a number of variables that are not currently included in the calculation of the flows of 'economic wealth' produced (in the GDP sense of the term). The variables in question may be greater or fewer in number and their inclusion is usually based on a concept of economic wealth that seeks to assess its contribution to well-being. These variables equate either to activities not included in the standard measurement of GDP (voluntary and domestic work, for example) but which contribute to well-being, to items that are accounted positively in GDP but do not contribute to well-being, or to various resources (notably natural ones) which, if improved or damaged in any way, impact on well-being.

The recent history of these attempts can be traced back to the work of Nordhaus and Tobin (1973) and to various other studies published in the 1970s that contributed to the academic debate among experts in national accounting.

The least that can be said is that such proposals for extending the boundaries of GDP have not met with universal approval. They have given rise to some important debates, which are outlined in this chapter.

1   Extending the boundaries of GDP: the controversies

Despite the academic and symbolic dominance of GDP, more or less successful attempts have been made to use the framework of national accounts in order to extend measures of economic wealth (production aspect) or of economic well-being (consumption aspect). National accountants themselves remain largely sceptical about these efforts, which change the very concept of gross domestic product in a fairly radical way. The arguments advanced by advocates of a 'reasonable' status quo include the 'commonsense' idea that the list of factors that contribute to well-being is potentially infinite. Once one gets involved in taking account of domestic and voluntary work or natural assets, there can be no knowing where to stop in extending the boundaries of national wealth as (re)defined in this way: leisure time, refreshing sleep, lovemaking... It's impossible to draw up national happiness accounts!

This type of commonsense argument does not stand up to historical analysis, and nobody is asking for happiness accounts. On the contrary, in fact, advocates of an alternative approach to wealth – among whom we number ourselves – explicitly state that the corollary of their demand for different kinds of accounts is the right not to count everything and to incorporate into assessments of progress various elements not based on indicators. On the other hand, history teaches us (Gadrey 2003) that the conventions used in national accounts to define the boundaries of the wealth deemed worthy of inclusion have changed over time in the wake of prolonged debates (cf. the inclusion of the activities of the civil and public services in the French national accounts, as outlined in Chapter VII). Thus there is no reason why they should not change again, in such a way that other, broader definitions of wealth and well-being can be evaluated alongside the market and monetary wealth that is currently taken into account. Simply maintaining the status quo is not 'reasonable' and has nothing to do with 'common sense'. On the contrary: it is conservative and an obstacle to innovation.

Efforts to construct an extended GDP would have to include an attempt to resolve the question of where to draw the boundaries, which would be tackled, as it always has been in the past, on the basis of conventions regarded as reasonable at the time. Thus neither refreshing sleep nor lovemaking would have to be included in GDP.

In fact, and contrary to the notion that the 'reasonable' position is to defend the status quo, it is our view that the strength of national accounting lies precisely in the fact that it is based on conventions (not in the sense of statistical conventions, whose presence nobody denies, but rather in the sense of the conventions used to define wealth and the scope thereof, which evolves over time). This is the key to its historic adaptability. Logical coherence is not an end in itself. The adoption of new conventions may lead to an increase in relevance, in terms of the public debate, that far outweighs what might be lost in terms of formal coherence.

We take the view that market (or quasi-market) GDP should be maintained as currently constituted, with the internal adaptations and revisions that are regularly made, as a crude measure of primary economic flows, including those that do not in any way contribute to well-being or that even reduce it. It is absolutely essential for certain analyses, including those relating to employment. However, we also think that it would be extremely useful to have at our disposal one or more synthetic indicators relating to an adjusted and enhanced GDP, which would have different objectives and be intended for use in different kinds of comparisons. It would bring into play value judgements, based on conventions, of the positive or negative contribution of certain human activities to well-being.

In general, the principle underlying these extensions or enhancements of GDP is the monetization of non-monetary contributions to well-being, whether those contributions be positive or negative (see Box IV.1).

2 An example of a convention: domestic work

Within the international circle of national accountants, there is now virtually no disagreement about the fact that the domestic production of *goods* should be recorded as production and included in GDP. The SNA 93 (System of National Accounts) nevertheless excludes the values of 'do-it-yourself' repairs and maintenance to vehicles or household durables, the cleaning of dwellings, the care and training of children or similar domestic or personal services produced for own final consumption. Clearly, this is a tricky point. After all, it is not easy for national accountants to explain to readers that, when

Box IV.1   Monetization in question

The monetization of certain non-market or non-monetary variables (voluntary and domestic work, natural resources, etc.) is a matter of dispute, among experts in national accounting as well as among charitable and voluntary organizations and academics. How can the problem be framed in general terms, before considering in each case what is actually possible or desirable?

One preliminary observation would be useful. It is aimed primarily at non-economist readers. What does it mean to 'monetize' one of the variables that make up a synthetic indicator? An example will serve to illustrate what is meant. Let us assume that it is agreed that voluntary activity adds to a nation's wealth (with wealth being defined broadly, as more or less equivalent to well-being) and that unemployment detracts from it, as does the destruction of ancient forests. It may be tempting, given these assumptions, to 'correct' or 'adjust' the traditional measures of national output or consumption for a given year (expressed in monetary terms, in billions of euros, for example) by supplementing them with an estimate of the monetary value of voluntary work over the course of the same year and deducting from them estimates of the 'social cost' of unemployment and of the decline in the value of the forests. True, it is no easy task to attribute a monetary value to such positive or negative contributions to well-being (various methods exist, each with its own particular conventions, and the results vary considerably depending on the method adopted). However, these attempts to extend GDP cannot be dismissed simply because these variables are, by their very nature, unamenable to monetization. In this case as in others, sufficient time has to be allowed for maturation and international exchanges. It may be that this path will turn out to be a blind alley; at the same time, however, such efforts may also take root, with academic and political alliances being forged at least for a sufficiently long period of time for it to be worth the trouble of ploughing the furrow in the first place. This is all the more likely to happen if attempts to extend GDP are accompanied by the development of other, less immediately economic indicators based on 'real' social and environmental variables.

(continued)

Box IV.1 *continued*

For our part, we are not enthused by the idea that, in order to make one's voice heard when advocating a not strictly economic vision of wealth and well-being that is, it is necessary to impute an economic value to all the non-economic variables. This could well be seen as a contradiction in terms that sets the seal on the triumph of economics as the supreme value and as the only credible justification for actions taken in defence of justice, social cohesion or the environment. After all, justifying voluntary work, an act of giving, and its contribution to society by imputing a monetary value to it, that is – whether one likes it or not – by reference to the market, is nothing less than an incredible admission of powerlessness in gaining acceptance for values other than those of the market economy! Nevertheless, it seems to us possible to overcome this initial rejection and to demonstrate greater pragmatism. Several arguments can be advanced in favour of such an approach, but the following is the most important one: money and the tools used for the purpose of monetization are not always tools of submission to economic values, to what Marx called 'the icy waters of selfish calculation'. Effective action in favour of social justice or the environment has long been based on economic and monetary tools, which can be used, for example, to 'constrain the selfishness' of those who 'pollute' society or the environment or to encourage them to behave less destructively.

The only real reservation one has to bear in mind with regard to the methods used to impute a monetary value to social and environmental variables seems to us to be the risk that these methods may be appropriated by experts, on the grounds of their (undoubted) complexity. If this happens, then the issues at stake become opaque for most actors who, without the means to make judgements, will be unable to assert their preferences on an informed basis. This risk will however be lower where these methods and indicators are not the only ones informing the public debate.

Mr Smith digs his vegetable plot or builds a garage and Mrs Smith does the housework, cooks the dinner and looks after the children or her elderly parents, it is only the former who is contributing to the wealth of the nation. People might well suspect that they are following an apparently technical argument that is concealing much deeper-seated conventions, and in particular conventions on the distribution of roles between men (seen as productive) and women (seen as unproductive).

However, before we align ourselves behind such accusations, which were made by feminists as early as the 1970s, we should examine the arguments advanced by national accountants, which are included among the underlying principles of SNA 93. These arguments can be divided into two main categories. The first are theoretical in nature and concern the potential exchangeability of these activities (see Box IV.3). They do not really stand up to close scrutiny. The second concern practical difficulties or political considerations and can, in our view, be overcome.

The pragmatic arguments advanced in favour of excluding domestic services from any evaluation of national output are essentially threefold. They concern, firstly, the difficulty of producing reliable annual data on the production of domestic services, secondly the difficulty of imputing a monetary value to such services and, thirdly, the allegedly diminished usefulness of such accounts 'for analysis and day-to-day economic policy' (Vanoli 2002, p. 307).

All this is interesting and merits consideration. Nevertheless, serious objections can be raised that tend rather to confirm us in the view that a gender convention is most definitely lurking behind all these arguments. After all, the first argument simply means that it is

---

Box IV.2   A new convention relating to the 'method of counting' and 'what really counts'

If an alternative convention on well-being and wealth is to become established, often by incorporating the previous one, various conditions have to be fulfilled. Whether or not these conditions are fulfilled will determine the strength of the alliances and networks that can be constructed around the new convention. Since we are dealing here with conventions that

(continued)

Box IV.2 *continued*

have to be translated into statistical procedures, one of these conditions concerns technical feasibility and reduction of the margins of uncertainty with regard to the definitions and results. In the case of the indicators outlined in this chapter and the next one, where a monetary value is imputed to variables that cannot be directly observed in the market or through production costs, one major challenge concerns the calculation of monetary equivalents for the value of certain services or variations (improvements or deteriorations) in a society's resources and inventories. This is undeniably trickier than measuring the output of the civil and public services (estimated on the basis of the sum of their production costs) or estimating fictitious rents (the value of the service rendered to its owner by a dwelling is calculated by imputing (fictitiously) rents to the consumption of households that own their own homes), not only because it is technically difficult but also because the 'margin of uncertainty' relates to the 'value systems' (what really counts?) on which the definitions of the equivalents are based. For example, should a monetary value be attributed to the 'nuisance' or 'damage' caused by economic growth on the basis of the repair costs incurred (and if so, which ones?), of the cost of preventing the damage (or the cost of adopting alternative solutions that do not cause such damage) or of so-called 'contingent' evaluations of consumers' propensity to pay in order to avoid the deterioration of a collective good (a classic problem with environmental externalities: see Box II.3).

Clearly, therefore, there are difficulties. However, if it is possible to introduce, at company or country level, rules of the 'polluter pays' type based on shaky conventions or to adopt highly debatable 'hedonic' conventions as a basis for calculating price indices and output volumes for computers and software and questionable conventions for measuring the output of banks and insurance companies (Gadrey 2002), then it is difficult to see why it would be impossible to agree on revisable conventions for estimating the value of voluntary work, domestic care or certain forms of environmental damage, even if the range of variables adopted had gradually to be extended as the international debate on methodology and conventions progresses.

Box IV.3   The potential exchangeability of self-produced goods and services

The inclusion of the domestic production of goods in GDP is a convention-based rule (SNA 93), 'which is based on the idea that goods are potentially more easily exchangeable than domestic services'. 'Exchangeable' has to be understood as meaning 'capable of being exchanged in a market'.

And yet domestic tasks, such as the care of children or the elderly, have long been exchanged in the market. According to some projections, more jobs will be created between now and 2010 in Europe in the group of occupations engaged in these last two activities than in any other occupation or group of occupations. One of the explanations for the growth in women's labour-market participation over the last fifty years is precisely the 'potential exchangeability' of the production of 'women's' domestic services. Without going so far as to include prostitution (which would lead, logically, to the inclusion of domestic sexual services, by virtue of their potential exchangeability), the least that can be said is that, in this example of the treatment of domestic work, rigour, consistency and theory have much less influence over accounting choices than the prevailing views of the social order. It is difficult not to interpret this kind of choice as an extension into the sphere of domestic activities, and on the basis of a gendered notion of wealth, of the old theory of unproductive labour, which was eventually abandoned in the so-called 'formal' sphere after two centuries, during which it dominated economic thinking.

not considered worthwhile making available the resources required to compile the required annual data. The second argument merely reflects the fact that the necessary energy has not been devoted to producing conventions for imputing a monetary value to services that are, in fact, considerably less difficult to define than financial or insurance services! As for the third argument, it amounts to nothing more than an admission that 'day-to-day economic policy' is concerned only with so-called 'socially organized activities' (Vanoli 2002, p. 307), in fact with activities associated with money and wage flows. It is as if the production of domestic services is entirely unconnected with a particular socially organized form of producing well-being, the scale of which depends on, among other things, family, fiscal and gender-equality policies.

Thus whatever argument is put forward in justification, it is because these activities are socially and politically undervalued (largely because they are traditionally carried out by women) that they are not monetized and included in national accounts, whereas non-market activities are related to the domestic production of goods, to private consumption and to own-account fixed capital formation.

For our part, we do not advocate the inclusion of these activities in the current market or quasi-market GDP, not because it is impossible but because the current GDP is useful, despite its carefully concealed imperfections, as a tool for analysing certain questions, namely those related to our market and monetary sub-system. We tend to favour the development of a new aggregate, which would have to be a matter for debate and which would, ideally, be accorded the same consideration, by both politicians and the media, as the current measure of GDP. It would be an indicator of national development, capable of taking account of both GDP and other measurable variables that impact strongly on our model of development and its sustainability. Examples of such aggregates will be provided in the following chapters. In methodological terms at least, the precursor to them all is Nordhaus and Tobin's 1973 paper.

3 Nordhaus & Tobin: the precursor of a new indicator

In economics as conceived by Walras, national accounting systems are supposed to be exempt from all ethical considerations, in accordance with the 'objectification' principle. Thus in the latest versions of the international systems (SNA 93 and ESA 95) it is clearly stated that illegal activities are not, in principle, excluded from the accounts.

It is this principle, which in reality is frequently infringed (Gadrey 2003), that Nordhaus and Tobin sought explicitly to challenge. In their famous 1973 article 'Is Growth Obsolete?', they attempt to identify elements which, in their view, do not contribute to 'economic well-being'. They named these elements 'regrettable goods'. The use of this term clearly indicates that normative, even moral value judgements are being made. In Box IV.4, we outline the basic principles of this pioneering attempt to construct a new measure of economic well-being or welfare.

Box IV.4   Nordhaus & Tobin's measurement of 'economic welfare'

Nordhaus & Tobin identify and calculate two indicators of 'adjusted final consumption', which they termed MEW, or Measure of Economic Welfare. The first of these is the 'actual MEW', which is obtained by adding to or subtracting from household consumption certain elements that make a positive or negative contribution to 'current economic welfare'. The second is the Sustainable Measure of Economic Welfare (SMEW), which in addition takes account of the imputed money value of certain stocks of economic, natural and human wealth (but not 'social' wealth: inequalities, economic insecurity etc. play no part here).

Measure of Actual Economic Welfare

The formula is as follows:

Actual economic welfare = personal consumption (as defined in national accounts)

minus: private instrumental expenditures: cost of commuting to work, some financial services ('regrettable'); education and health purchases (yield as intermediate services and included as such in the SMEW); durable goods purchases (replaced by the estimated value of the services rendered by the stock of durable goods in households); disamenities of urbanization (compared with housing in rural areas)

plus: imputation for the services rendered by durable goods; imputation for leisure; imputation for productive non-market work; a (small) proportion of public expenditure regarded as contributing to current well-being (postal services in particular); estimate value of services rendered to individual by the stock of public capital.

(continued)

Box IV.4 *continued*

Sustainable Economic Welfare

This measure (we will not go into the details of it here) is based on an evaluation of the variations in a stock of public and private wealth that contains four components:

a) net reproducible productive capital (equipment, infrastructure, etc.);
b) non-reproducible capital, confined here to the value of land and net foreign assets;
c) educational capital, the value of which is estimated on the basis of the number and average cost of the years of education undergone by economically active individuals;
d) health capital: the cumulative value of public and private expenditure on health, to which a depreciation coefficient of 20% per annum is assigned.

Let us examine some of these elements in the light of the main objections raised during the debates on Nordhaus and Tobin's proposals during the 1970s and subsequently, starting with the deductions and additions that figure in the definition of the actual MEW.

Nordhaus and Tobin deduct all or part of collective consumption (public expenditure) on the ground that it does not necessarily contribute to *current* welfare. They advance three arguments in justification of this position. Firstly, part of public expenditure must be regarded as investment (and reclassified as such) and not as consumption; this applies to education and public health services, as well as to some services that are consumed directly by individuals (transport and housing, for example). Another part of public expenditure also has to be subtracted, namely the 'regrettable' elements, a notion that they introduce in a fairly restrictive way but that was to be widely used and extended subsequently. These regrettable elements equate to final expenditures made for reasons of security, prestige or diplomacy, 'which, in our view, do not directly increase households' economic welfare'.

(continued)

Box IV.4   *continued*

For Nordhaus and Tobin, the largest 'regrettable' item is defence expenditure. Other national accountants have objected to this position, arguing that defence expenditure can be regarded as providing the direct satisfaction of peace and security. This is a classic example of the conflicts that can arise when attempts are made to define conventions on what constitutes wealth and welfare.

Some items of public expenditure on goods and services regarded as intermediate are also subtracted from final production. This applies, for example, to direct services to firms provided by local and central government, as well as to the expenditure incurred in maintaining a sound and wholesome health and social environment.

Household consumption is also adjusted downwards through the exclusion of those elements that do not contribute to welfare. Thus certain items of work-related expenditure (e.g. the costs of commuting to work) are deducted from private consumption, as are all or part of medical and education expenditure (which is regarded as gross investment) and all expenditure (on housing and other items) that could be considered as an investment.

Nordhaus and Tobin also deduct the negative externalities linked to urbanization and the ensuing congestion, in accordance with the principle that part of the differential between the pay of city dwellers and that of their rural counterparts is intended to compensate for the disadvantages of working in urban areas. They were not the first to do this. As early as 1949, Kuznets pointed to the 'inflated costs of urban civilisation'.[1] Other national accountants object that there are also advantages associated with urban living, as well as disadvantages associated with rural living, of which no more account is taken when calculating GDP than of the negative urban externalities cited by Nordhaus & Tobin. Once again, we are engaged here in a debate that relates directly to the norms that define the quality of life in society.

The measure of national welfare obtained on this basis differs very considerably from that based on household consumption, as measured by the standard accounts, largely because of the enormous and certainly excessive (monetary) value attributed to leisure and non-market activities. On the other hand, environmental considerations play very little part. An even more surprising result is that the variations over time in economic growth and in these indicators of welfare are strongly correlated over the period 1929–65, which leaves the authors to conclude that growth is not, after all, such a bad indicator of welfare. It is this type of finding that recent studies conducted on a similar methodological basis have begun seriously to call into question (see next chapter).

V
Environmental GDPs and the Ecological Footprint

This chapter is given over to predominantly environmental indicators that have been proposed as alternatives to GDP and, like national accounts, use a homogeneous unit of account. This is usually money. The result is an environmental GDP, in the sense that monetary estimates of the value of gains or losses in environmental quality, together with certain social characteristics, are added to the figures normally included in national accounts. The ecological footprint indicator, which is outlined in this chapter (§4), is an interesting exception.

Experts in national accounting continue to debate the question of how best to incorporate environmental considerations into their accounts. We do not investigate their current controversies in any detail in this book. As far as international institutions are concerned, one reference work, and a source of useful data, is the United Nations' impressive manual, entitled *The Handbook of National Accounting* (2003). This can be supplemented by the World Bank's studies on 'genuine savings', which are significantly less detailed and which we will outline briefly in §3.

1 Green national product and the index of sustainable economic welfare (ISEW)

Most of the indicators outlined in this chapter are concerned with the 'green national product' and various 'indices of sustainable economic welfare' (ISEW), as well as the GPI or Genuine Progress Indicator. To the best of our knowledge, the first internationally cited version of the ISEW was published in an annex to the important book written by John Cobb and Herman Daly (1989). However, it is the book edited in 1994 by Clifford Cobb and John Cobb that marks a watershed. Since then, there has been a burgeoning of initiatives in a number of countries, including Canada, Germany, the UK, Austria, the Netherlands and Sweden. In some of these cases, the research was carried out in state-funded environmental institutes. The paper by Jackson & Stymne (1996), published by the Stockholm Environment Institute and available on line, gives a good overview of these initiatives.

Another good example is provided by the Friends of the Earth's index of sustainable well-being. Since 2001, this international NGO, working in cooperation with the New Economic Foundation (an alternative think-tank specializing in social reporting) and the Centre for Environmental Strategy at the University of Sussex, has been developing its own ISEW for the UK. It also offers an on-line tool that enables anyone 'to create their own ISEW' by weighting the variables differently and sending their final choices to the FoE researchers so that they can construct a variant index based on this 'electronic poll'. This index, like those just cited above, is fairly meticulous on the methodological level, although here too enormous uncertainties remain. The novel features that set it apart from its predecessors include three revisions relating to the treatment of income inequalities, evaluation of the damage associated with global warming, and the cost of destroying the ozone layer.

Box V.1 lists the variables included in such indicators, as well as a broad outline of the methods used to impute a monetary value to them. Although the weightings vary, all these indicators combine contributions to sustainable well-being from several different areas: economic (standard of living), social (e.g. inequalities) and environmental, as well as the contributions of non-monetary activities to individual well-being (e.g. value of household work).

Box V.1   The main variables and methods used in these indices (after Jackson & Stymme 1996)

Table 4 summarizes the variables used to construct indicators of the ISEW type and provides some indications of methodology, some of which are discussed subsequently. Jackson and Stymme's study deals with the subject at some length and is very explicit

(continued)

Box V.1 *continued*

about these methods and their limitations. For a highly technical debate on these methods, readers are referred to Cobb & Cobb (1994) and C. Cobb et al. (1995).

In addition to those already mentioned relating to the classification of activities or expenditure as defensive or non-defensive, the many uncertainties surrounding the construction of these indices also include those relating to the imputation of a monetary value to various kinds of environmental cost and damage. There are three categories of cost or damage:

a) the decline in the 'sustainable value' of land (whether cultivated or not);
b) the loss of non-renewable natural resources (mainly fossil fuels); and
c) long-term environmental damage, primarily that linked to Co_2 and CFC emissions (ozone layer).

In all these cases, conventions and very approximate estimates are used, for want of any better solution.

Let us take the particularly important example of the estimation of the costs associated with the irreversible decline in fossil fuels such as oil and natural gas. One of the most frequently used options is to use replacement costs as a means of attributing a monetary value to this decline. This involves estimating the current cost of replacing each 'barrel-of-oil equivalent' consumed (on the basis of non-renewable resources) by renewable energy resources. This is interesting, and it is normal for this replacement cost to vary with the prices of the resources (whether renewable or not), since this is the price of halting the decline in 'sustainable well-being'. However, the issue becomes more complicated if it is assumed, which it is also reasonable to do, that earlier use of fossil fuel resources have also contributed to the reduction in well-being: it is not possible simply to perform the calculation from the present-day onwards. The whole process has to be 'updated' by performing the calculation over a long period and choosing a discount factor, which is what the authors of these indices do. However, these are conventions that are difficult to define, particularly since technical progress cannot be forecast.

Table 4   The variables used in the ISEW and their monetization

Variables	1992 value (Sweden*)	Indications of methodology
Consumer expenditure	502	Basis for the indicator
Income distribution–	0.77**	Use of Gini coefficients to weight consumption level
Weighted personal consumption	653 (= 502/0.77)	
Services: household labour	+242	Imputation of monetary value on basis of hourly rate of pay for cleaner
Difference between expenditure on and services from consumer durables	–36	Evaluation based on investment accounting methods
Public expenditures on health and education	+33	Only the share of expenditure regarded as non-defensive is included (convention: half of such expenditure)
Defensive private expenditures on health and education	–10	Deducted in part (50% regarded as defensive)
Costs of commuting	–37	Deducted (regarded as defensive expenditure), on the basis of studies of distances travelled and average cost
Costs of car accidents	–8	Deducted (regarded as defensive expenditure), on the basis of studies relating to particular years
Costs of water pollution	–24	Deducted, on the basis of data on the evolution of river pollution and American attempts to assess the costs of water pollution
Costs of air pollution	–24	Deducted, on the basis of statistics on SO_2, NO_x, CO and particulate emissions and estimates of the cost of damage.

Table 4   The variables used in the ISEW and their monetization – *continued*

Variables	1992 value (Sweden*)	Indications of methodology
Costs of noise pollution due to car traffic	–8	Deducted, on the basis of a Swedish estimate of the average annual cost of per person exposed to noise in excess of 55 decibels and of the number of such individuals.
Costs of loss of wetlands	–2.4	Estimation of losses of natural capital
Costs of loss of farmlands as a result of urbanization or the non-natural decline in soil quality	–3.4	Estimation of loss of farmlands in both area and value terms, on the basis of an adjusted price per hectare that takes account of variations in soil quality.
Depletion of non-renewable resources	–166	Calculation on the basis of replacement cost of non-renewable energy resources (see comments above)
Long-term environmental damage (incl. CO_2 emissions and nuclear waste)	–73	Notion of 'environmental debt', calculated on the basis of the estimated environmental cost of each barrel-of-oil equivalent consumed from non-renewable energy sources.
Ozone depletion costs	–22	Application of a unit cost for each kilogram of cumulative world production of CFC.
Net capital growth	+60	Adoption of a variant of the standard measure of the increase in net productive capital in the market sphere
Change in the net international position	+49	A country's indebtedness is regarded as damaging to sustainable development (burden on future generations)
ISEW (1992)	595	Sum of preceding lines

(*) in billions of SEK (Swedish crowns) in 1985. In this column, the + and – signs indicate what is added or deducted from final consumption.

(**) base 1 in 1950 (considerable reduction in inequalities between 1950 and 1992)

Defensive expenditure

One of the difficulties encountered in constructing these indicators concerns so-called 'defensive' expenditure. The idea is a simple one (it was mentioned in Chapter II when the production of 'compensatory' goods and services was discussed), although it is tricky to implement; it involves refusing to include as a positive contribution to 'real' wealth any expenditure (and the corresponding production) that serves essentially to repair the 'harm' or 'collateral damage' caused by our growth model and lifestyles. The difficult cases include the treatment of certain items of expenditure depending on whether or not they are regarded as defensive. For example, advocates of these indicators generally adopt the convention that half of public expenditure (and of private household expenditure as well) on education and health is defensive in nature, which is debatable, although some of the arguments advanced merit consideration. After all, a proportion of education expenditure does indeed help to improve the relative position of certain individuals in the labour market without any overall increase in competences and productivity, while a proportion of health expenditure is incurred in the treatment of conditions linked to environmental degradation, social inequalities, working conditions and various accidents associated with economic activity, as well to weak prevention policies. Taken overall, these studies have a marked tendency to regard most public expenditure as defensive, which seems to us highly debatable. For example, the fact that the level of health expenditure in France rose from 3 to 9.6 per cent of GDP between 1950 and 2003 can probably be explained more by the 'increasing cost' of the spectacular rise in life expectancy from 68 in 1954 to 79 in 2003 than by any explosion in expenditure on making good the damage to human health caused by our growth model or by waste (although the problem of excess consumption of medication is a very real one). On the other hand, however, account has to be taken of recent and convergent estimates of the destructive effects of an unhealthy environment. According to the World Health Organization (cited in the *Lancet* of 19 June 2004), the poor quality of the environment may be responsible for the death of one child in three in the world.

Calculating the ISEW

The indicator of sustainable economic well-being can be expressed approximately by the following formula:

ISEW = personal consumption expenditures (base or starting point for the calculation) + value of unpaid household services + non-defensive public expenditures – private defensive expenditures – losses due to deteriorating environment – natural non-renewable resource depletion + net capital growth.

This formula is generally corrected by making two adjustments. The first concerns variations in income inequalities. A weighting coefficient (the Gini index) reflecting the evolution of these inequalities is applied to total final consumption. The second adjustment relates to household durable goods. As can be done with productive investment, a distinction is made between the purchase value of these goods and the value of the services they provide.

It will be noted that the above formula (specified in greater detail in Table 4 on pages 66 to 67) does not include two variables which are, nevertheless, generally considered to contribute to well-being (by Nordhaus and Tobin, for example), namely leisure time and human (or educational) capital. The authors put forward various theoretical and practical arguments to justify this.

Should these indicators be rejected while we wait for more reliable estimates on which international agreement can be reached? This is not our view, for several reasons. Firstly, even with an enormous margin of uncertainty, as high as 50 per cent for some of the estimated environmental costs (which is certainly too high), the scale of the losses of sustainable well-being associated with these phenomena remains very considerable. For this reason, they cannot be ignored. Secondly, as soon as these indicators are used for diagnostic purposes over very long periods (one or more decades), even such a large margin of error is considerably reduced and has little impact on major trends, which are the ones that count as far as sustainable development is concerned. Finally, over and above the synthetic indicator, there is a need – as there always has been – to analyse its various components, both in (imputed) monetary terms and in 'real' or physical terms. This might produce more reliable estimates.

Despite its limitations, however, the synthetic indicator remains a useful reference point by providing a rough order of magnitude. This is all the more the case since some of the trends that it points to are greater than the margins of error in its construction. These observations also apply to the ecological footprint, another indicator that is as important as it is uncertain (see below, page 74).

Two illustrative graphs

The two graphs reproduced below and opposite (after Jackson & Stymne 1996), which relate to the UK and Sweden, are interesting in at least two respects. Firstly, they can be interpreted in terms of the variations over time in each country. Such a reading would highlight the marked decline in 'sustainable well-being per person' in the UK between 1974 and 1990. Such decline is much less perceptible in Sweden. The authors of this research show that the causes of this divergence in trends include, among other things, the pronounced increase in social inequalities in the UK, which is the converse of the trend in Sweden over the same period. A second interpretation of these graphs would involve examining the gap between the two curves, that is the difference in absolute terms

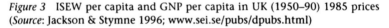

Figure 3   ISEW per capita and GNP per capita in UK (1950–90) 1985 prices (*Source*: Jackson & Stymne 1996; www.sei.se/pubs/dpubs.html)

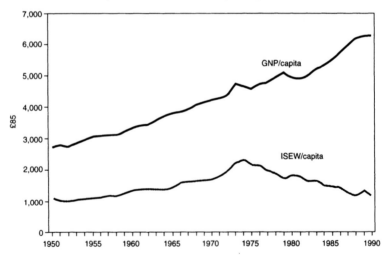

Figure 4 ISEW per capita and GNP per capita in Sweden (1950–92), 1985 prices (*Source*: Jackson & Stymne 1996; www.sei.se/pubs/dpubs.html)

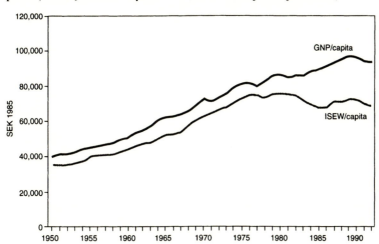

between GDP (in fact GNP, which is a similar parameter) and the 'corrected' index. As early as 1950, this gap is much more pronounced in the UK; it also widens markedly in this country and the difference between the two situations is such that it cannot be attributed simply to statistical uncertainties.

2 The Genuine Progress Indicator (GPI)

This indicator, which is now well known in the USA and is very similar in its inspiration and methodology to the indices of sustainable well-being outlined above, was developed by researchers at Redefining Progress, which is a non-profit public-policy organization set up in 1994. The GPI was disseminated from 1995 onwards. Research institutes in several countries, including Germany, the UK, Canada and Australia, were quick to adopt the index and to apply it to their own countries. The following elements are reproduced below:

- Firstly, in Table 5, the 'GPI accounts' for 2000, which show how the move from total household consumption to the GPI is effected by adding or subtracting a number of variables to which a monetary value has been attributed. The list of these variables

Table 5   American GPI in 2000 (in billions of dollars at 1996 value)

Personal consumption	5 153
Economic adjustment	
Income distribution	–959
Net foreign Lending or borrowing	–324
Cost of consumer durables	–896
Social adjustments (costs)	
Cost of crime	–30
Cost of car accidents	–158
Cost of commuting	–455
Cost of family breakdown	–63
Loss of leisure time	–336
Cost of underemployment	–115
Environmental adjustments (costs)	
Cost of household pollution abatement	–14
Cost of water pollution	–53
Costs of air pollution	–39
Cost of noise pollution	–16
Loss of wetlands	–412
Loss of farmlands	–171
Depletion of non-renewable resources	–1 497
Costs of long-term environmental damage (environmental debt)	–1 179
Cost of ozone depletion	–313
Loss of old-growth forests	–90
Benefits added to the GPI	
Value of household and parenting	+2 079
Value of volunteer work	+97
Services of consumer durables	+744
Services of highways and streets	+96
Net capital investment	+476
GPI	**2 630**

(*Source*: C. Cobb et al. 2001)

is very similar to the one that appears in the table in the previous
text box. However, it does include variables that were not taken
into account in the Swedish case, generally because the relevant
data were not available;
• Secondly, a graph (Figure 5) that compares the variations in GPI
and GDP per capita in the USA from 1950 to 2000.

Figure 5 GDP and GPI per capita, United States, 1950–2000, in dollars at 1996 value

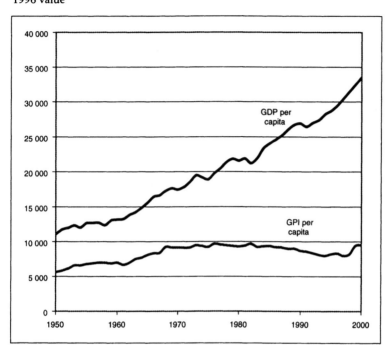

3 The World Bank's genuine savings indicator

This monetized indicator, which could also be described as an indicator of increases or decreases in genuine wealth, seeks to contribute to the synthetic measurement of a country's sustainable development by adjusting the standard definition of national savings by adding or subtracting various non-economic resources, notably environmental resources. It does not include any social variables. It has been very strongly criticized by Everett and Wilks (1999), who quote Joan Martinez-Alier, the Spanish economist and environmental specialist.

The principle for calculating this genuine savings indicator (subsequently renamed 'adjusted net savings') can be summarized by the following equation (Hamilton 2001):

Genuine savings = net savings (i.e. gross domestic savings –
consumption of fixed capital) + education
expenditures – (energy depletion, mineral
depletion, net forest depletion, carbon-dioxide
damages)

Monetary evaluation of the depletion of non-renewable mineral
resources is based on the notion of supplementary net income,
which measures the difference between sales prices after extraction
and the economic costs of that extraction (including prospecting).
This is also what is called resource rent. It remunerates both the
owner of the resources (mines, wells, etc.) and the extractors (Vanoli
2002). In the case of CO_2 emissions, the method adopted involves
estimating the marginal social cost of each metric tonne of CO_2
emitted. Forest depletion is valued as the stumpage value (price –
average logging cost) of the volume of commercial timber and fuel-
wood harvested in excess of natural growth in commercially valuable wood mass of the year (Hamilton & Clemens 1999, p. 340).

4 The ecological footprint

This indicator is the only one of those examined in this book to be
purely environmental, with an anthropocentric view of nature (that
is one that focuses on the relations between man and nature).

The ecological footprint of human activity

At international level, two NGOs working in close cooperation with
each other have been attempting since the second half of the 1990s
to popularize the use of a synthetic indicator, the ecological footprint, that very vividly reflects the extent to which human beings
use nature for the purposes of material production and consumption. They are the organization Redefining Progress, already mentioned in connection with the GPI (§1), and the World Wildlife
Fund (WWF), which was renamed the World Wide Fund for Nature
in 1987. Their work has its origin in a concept developed and
applied in a joint publication (1995) by two researchers at the
University of British Columbia in Vancouver, Mathis Wackernagel
and William Rees. This indicator is still little known, but is gaining
in influence.

The idea underlying its construction can be summarized as follows. Human activities related to production and consumption use natural resources, some of which are non-renewable (oil and natural gas, mineral deposits) whereas others are renewable, in the sense that they can reproduce or regenerate without human intervention: soil, forests, water, atmosphere, climate, animal species that reproduce naturally (fish, etc.). The ecological footprint focuses solely on these latter, since its advocates take the view that it is they that pose the most serious problems in the long term.

These resources have two functions with regard to human activity: to provide inputs (raw materials) for production and consumption and to absorb (recycle) waste, including emissions of CO_2 and other gases into the atmosphere. One of the question that arises here is whether or not these natural resources, which in theory are renewable, are today being depleted because the volumes being used exceed global capacities for renewal. The ecological footprint approach offers an opportunity to draw up a balance sheet of this kind, using a newly developed unit of account that synthesizes partial balance sheets drawn up for various different categories of natural resources that do not share a common unit of account, such as those found, for example, in Lester Brown's remarkable book (2001).

Advocates of the ecological footprint can be said to have adopted a perspective similar to that of the pioneers of national accounting but with the aim of drawing up a 'budget' for the natural world (in its relations with human activity) that reflects mankind's 'ecological debt' (when mankind's borrowings exceed the world's natural capacity to regenerate resources).

The method

The principle is a simple one: virtually all the renewable resources used to satisfy a human community's economic needs can be translated into land and water areas that constitute the 'footprint' of human activity. The value of such a conversion can be illustrated by the following analogy: the tribes that lived from hunting and gathering could not survive without the resources of a natural territory, the extent of which depended on the tribe's consumption (that is on the number of people in the tribe) and on the territory's bio-productive capacities (bio-capacity), that is its ability to regenerate the resources used by the hunters and gatherers. They could not allow themselves to exhaust the territory's water, vegetable or animal resources, which are certainly reproducible *but only within certain limits and in accordance with the natural cycle.* With their experience of ecological crises, contemporary economies are now rediscovering that they too are subject to certain laws related to the finite nature of natural resources and the limits of regeneration. Our territory – the planet – is probably being exploited by modern tribes beyond its capacities for reproduction. Thus it is the difference between the area of the planet exploited by modern tribes (the footprint) and the area available for such exploitation that this method seeks to estimate.

More specifically, the starting-point for the calculations is a given population's final consumption (that of a country, for example), which is then converted into a number of 'footprints' representing the area exploited for each item of consumption on the basis of the technologies currently used in production and in exploiting natural resources (see Box V.2 opposite).

The main types of natural areas used are listed in the first column of Table 6. It is essential to point out that the various areas of the earth's surface that have to be exploited in order to achieve a given level and mode of consumption in a given country do not have any boundaries. This is a major difference from the case of the primitive tribes cited above. As soon as the consumption in a given country or community begins to include goods produced from natural resources that come in part from outside or the waste produced has to be recycled in the natural environment outside the territory analysed (the most important example being that of CO_2 emissions), the only way

Table 6   The ecological footprint of the United States in 1999 in millions of hectares

Crop land footprint	415
Grazing land footprint	89.7
Forest footprint	358.9
Fishing ground footprint	86.9
Built-up land	103.7
Energy	1 665.6
of which nuclear	(140.2)
Total	2 719.9

Source: WWF Living Planet Report 2002 (online)

of calculating footprints is in terms of the surface of the world as a whole. The CO_2 emitted in one country is not all absorbed by the forests in that same country. This is the reason why a country's ecological footprint is calculated on the basis not of the bio-capacity of its own natural resources but in terms of a global unit of account: the global hectare or average bio-productive hectare (see Box V.2). That said, however, the bio-capacity of an individual country or territory is not excluded from the accounts that are drawn up, which makes it possible to compare the areas that country exploits on a global scale with its own stock of bio-productive areas. For example, William Rees has estimated that, just to produce food, the Netherlands use ('import') an area of approximately 100,000 km² spread across the world, mainly in developing countries, that is five to seven times the area of productive land within its own borders. On an economic level, this country has positive trade balances. On the ecological level, it has enormous deficits with the rest of the world, but for the moment these deficits are invisible. It is certainly normal and desirable, in an urban civilization, that a territory with a high population density should have a permanent ecological deficit. Firstly, however, it is no bad thing for its population to be aware of the extent of this deficit (which indicates the extent to which the country is dependent on natural resources located elsewhere) and of its evolution. Secondly, things become more worrying when the global total of these deficits puts at risk the renewal of the natural resources imported by each of the countries with a large footprint.

An ecological footprint can be calculated for the whole of humanity, for a country, for a region or town, for a household (on the basis of what it consumes), for an item of final consumption (food, housing, transport), and so on.

Box V.2   The translation into global hectares of a population's consumption

The principle of this translation can be summarized as follows (for further details of the methodology and results, see the WWF reports and Wackernagel & Rees [1995] and Wackernagel & Onisto [1997]). Six types of natural area are exploited in order to meet a population's consumption needs (and the corresponding production). It should be noted that these areas are calculated on

(continued)

Box V.2   *continued*

the basis of the world average bio-productivity of each major natural resource: soil, forests, oceans, etc.

1. The cropland footprint is the area required for growing agricultural products for human and animal consumption or for industrial production (cotton, jute, rubber etc.).
2. The grazing land footprint of a population corresponds to its consumption of meat, dairy products, hides and wool that come from livestock that are not crop-fed but occupy permanent pastures.
3. The forest footprint of a population is the area required to produce the forest products it consumes. This includes all timber products. Wood or charcoal burnt as fuel are included in the energy footprint.
4. A population's fishing ground footprint is the area required to produce the fish and seafood it consumes. Account is taken of the fact that not all fish species are equal in terms of their requirements for biological productivity (ocean's primary production).
5. Built-up land is the area required to accommodate housing, transport, industrial production and hydroelectric power.
6. A population's energy footprint is the area required to produce the energy it consumes. Four types of energy are taken into account: energy from fossil fuels (coal, oil and natural gas), biomass energy (wood and timber), nuclear energy and hydropower. The footprint of fossil fuel combustion is the area of forests that would be required to absorb the resulting carbon dioxide emissions. The biomass footprint is the area of forests needed to grow the biomass. Nuclear energy is included in the energy footprint and entered in the accounts on the basis of the highly debatable convention that it is equivalent to fossil fuel per unit of energy (excluding nuclear power would reduced the world energy footprint by less than 4 per cent). The hydropower footprint is the area occupied by hydroelectric dams and reservoirs.

Figure 6 World ecological footprint 1961–99 (after Living Planet Report, WWF 2004 online)

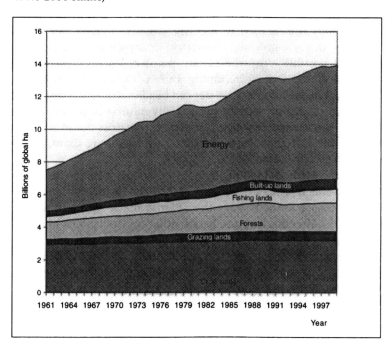

An invisible debt

According to the WWF reports, the world ecological footprint has increased considerably since 1960, rising from 70 per cent of the Earth's surface that can be used for productive purposes in 1961 to 120 per cent in 1999 (see Figure 6).

If this latter figure is correct, it means that mankind is currently borrowing a volume of natural resources each year that is 20 per cent greater than the annual flows of regenerated resources the natural world can produce. However, this deficit is invisible, for four reasons. Firstly, this form of account remains virtually unknown. When it comes to issues of this kind, what is not counted simply does not count. Secondly, even in the economic sphere, a high level of indebtedness does not necessarily have visible consequences in the short term. And when structural indebtedness remains hidden, only serious crises propel it into public view. Thirdly, since ecological assessments are dominated by economic evaluations, prices are the only means used to identify possible resource depletion. However, this is a misleading indicator, since technological progress makes it possible – up to a certain point at least – to extract or exploit resources that are becoming scarce or exhausted without any significant increase in price. Water can be pumped from water tables at increasing depths without causing any upsurge in prices as long as there is water to be pumped. However, that does not in any way change the fact that water resources are becoming dramatically scarce in certain countries.

Even the methods used to attribute a monetary value to the losses of renewable resources in order that they can be incorporated into indicators such as the ISEW or the GPI may not be sufficient to raise the alarm. After all, they are based either on the current prices of these resources (when a price exists, as it does in the case of losses of cultivated land, for example) or, more commonly, on the estimated costs of the damage caused by these losses of resources. However, in the case of water and of other resources as well, there is every likelihood that damage in future will be more serious than the damage that we are now beginning to observe and measure. This is why a physical accounting system intended to provide early warnings of damage based on the actual state of the resources themselves rather than on estimated monetary values, which seem to be ill-suited to this type of crisis forecasting, could be so valuable.

Finally, the fourth reason for the lack of attention being paid to the current depletion of renewable natural resources is that its negative consequences for daily life are not (yet) affecting the dominant economic, political and media players or the privileged classes. These groups have by far the largest 'footprint' but, for the moment at least, they also have the means to shift the burden of these consequences on to others, to protect their environmental surpluses and, increasingly, to privatize the services provided by eco-systems that were previously accessible to all as common goods.

We have already noted that this system of accounting is based on the *currently prevailing* modes of consumption and production technologies. This is a crucial point: other, non-regressive lifestyles and different production technologies (e.g. renewable energy sources, hydrogen power – if the faith some people are placing in this technology is confirmed – and agricultural methods that make less demand on water tables and do not exhaust the natural capacities of the land, etc.) may reduce the ecological footprint very considerably without impacting adversely on what are regarded as the fundamental objectives of civilization in terms of the quality and diversity of foodstuffs, housing, travel, medical care, etc. Such calculations, whose gloom-mongering applies only to our current growth model, do not seem to lead ineluctably either to a reversal of economic growth or to demographic Malthusianism or even back to the 'good old days'. However, the debate is ongoing and is of vital importance (see Box V.3).

Box V.3 The negative growth of what?

A debate has recently got under way in several European countries on 'negative growth', which its advocates advance as the only way of avoiding ecological catastrophes in the long term. This debate is focused to an excessive degree on the admittedly very real need to scale back activities that are in effect environmental weapons of mass destruction. The harm they cause results not only from their emissions of greenhouse and other gases that are harmful to the environment and to health, as well as of particles that cause various diseases, but also from the pressure they exert on non-renewable resources and even on some so-called renewable resources, which could themselves become scarce if exploited to excess. Such activities must either be scaled back or become the locus of innovations in the way they are produced and consumed that would make the pressure they exert on the environment sustainable.

However, it is no less important to draw up a list of the very many activities that combine low environmental pressure with a strong contribution to individual and collective well-being (this is where the value of alternative indicators in such diagnoses becomes evident). It is here that the keys to a form of 'growth in activity' that would not threaten the environment, social cohesion or employment lie. However, even though attention must also be paid to agriculture, manufacturing industry and

(continued)

Box V.3   *continued*

constrution (because they meet essential needs), the vast majority of these activities and these 'reservoirs of employment and
of sustainable human development' are *services*, whether private,
public or voluntary. A blanket invocation of negative growth
is scarcely 'sustainable' in view of the inadequacy in most
developed countries of housing provision and home-help
services for dependent elderly people and the handicapped,
care facilities for young children, many aspects of health care,
education, training and research, public interest activities,
social and cultural services, public transport, energy distribution, alternative distribution channels, and so on. To say
nothing of the major tangible and intangible investments in
sustainable human development which, in this scenario, could
provide the basis for a Keynesian policy aimed at sustainable
economic revival and the creation of good-quality jobs. Most of
the activities that could give a different meaning to growth are
relational and professional services provided at local level, that
is precisely the activities that, for the last 25 years, have been
responsible for the vast majority of the new jobs that have been
created. Their ecological footprint is remarkably modest.

Pronounced inequalities in the footprint

For the organization Redefining Progress, the ecological footprint is
a concept that closely links the notions of sustainable development
and equitable development. The requirement for equity comes into
play in three ways. Firstly, of course, there is intergenerational
equity, which is the basis of the notion of sustainability. The ecological debt will have to be repaid in one way or another by future
generations. Secondly, there is national and international equity,
since the statistics reveal the existence of enormous inequalities
between both countries and social groups in terms of ecological
footprint (see Figure 7). In 1999, the average ecological footprint of
an inhabitant of North America was, according to the WWF, 9.6
hectares (which is five times the 'bio-productive' area per person in
the whole world, put at about 1.9 hectares), compared with less

Figure 7 Ecological footprint, by region and income group, 2001 (after *Living Planet Report,* WWF 2004 online)

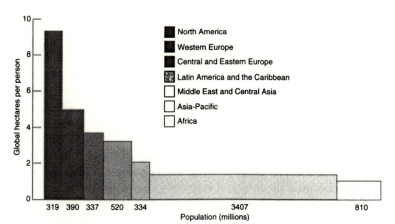

than 1.4 for an inhabitant of Africa or Asia. These figures show that the equivalent of five planet Earths would be needed if all the worlds' inhabitants were to be guaranteed an American lifestyle over the long term and on the basis of current production technologies! And almost three would be needed if the French lifestyle were the point of reference. Finally, the inequalities between social groups within the same country are no less pronounced: the differences can range from a factor of 1 to a factor of 10, and would be significantly greater if the most luxurious lifestyles were taken into account.

5 The Dashboard of Sustainable Development

We have excluded from this book any examination of indicators that are concerned *solely* with the environment, with one exception (the ecological footprint). The only significant attempt to construct a (non-monetary) synthetic indicator with a high environmental component that we will mention is the Dashboard of Sustainable Development developed by the Consultative Group on Sustainable Development Indicators (CGSDI), which has been coordinated since its foundation in 1996 by the International Institute for Sustainable Development in Winnipeg, Canada.

This dashboard contains 46 indicators in three main areas (environment, economy and society) and includes no fewer than 100 countries! It encompasses the following aspects and variables:

- environment (13 indicators): water, air and soil quality, levels of toxic waste;
- economy (15 indicators): GDP, investment, productivity, competitiveness, inflation, energy consumption;
- society (18 indicators): crime, health, poverty, unemployment, education, governance, military expenditure and cooperation.

This tool, which is not primarily an indicator, combines a free, non-commercial computer program with an international database that can be used flexibly and enables users, whether individuals, associations or pressure groups, to consult a wide range of data, to build up their own customized dashboards and possibly in the future to construct one or more national synthetic indicators by varying the number and weighting of the variables as they wish. Box V.4 discusses the problems involved.

Box V.4   Debatable method and choices

The data used to construct each of the Dashboard's indicators are available on a scale ranging from 0 to 1000. The country with the highest absolute value is allocated 1,000 points, while the one with the lowest absolute value is given a score of 0 points. One of this Dashboard's novel aspects is that the weighting of each area can be modified in a user-friendly way depending on users' specific interests and objectives.

However, considerable problems remain, and they probably undermine this tool's international credibility. For example:

- The Dashboard is currently dedicated solely to international comparisons at a given moment in time. It cannot be used to track performance over time.
- This Dashboard attaches considerable (too much?) importance to the economy: tariffs, integration into the global economy, trade, regulatory barriers, labour costs, all from an ambiguous point of view (fairly often free trade).

(continued)

Box V.4 *continued*

• The parameters adopted in order to construct these indicators are sometimes problematic. For example, the rate of poverty used is defined on the basis of the percentage of the population living on less than 1 dollar a day, in purchasing power parities, which makes it impossible to differentiate between countries on the basis of relative poverty thresholds, which are significant in some cases. Thus all developed countries, from Sweden to the USA, are given a score of 1,000!

• The designers of the Dashboard were aware of possible problems with conflicting judgements with regard to certain indicators. However, they resolved these conflicts by prejudging them. For example, in assessing overall 'performance', should the rate of urbanization be regarded as a positive or negative determinant? This issue is resolved by deciding that it is a negative indicator. Thus Belgium, whose rate of urbanization is 97 per cent, receives a score of 0 while Albania, where the rate is of the order of 41 per cent, receives a score of 1,000. Another example is the question of labour costs. Do high labour costs always have to be regarded as a negative factor? Once again, the Dashboard prejudges the matter, giving Norway the lamentable score of 36 because of the high cost of labour in its manufacturing industry, whereas Poland is given a score of 1,000 by virtue of its low labour costs!

VI
The Index of Economic Well-Being

Lars Osberg, of Dalhousie University, in Nova Scotia, carried out his work on 'economic well-being' in Canada in the mid-1980s, but it was not until 1998 that, working in collaboration with Andrew Sharpe, of the Center for the Study of Living Standards in Ottawa, he published series for Canada and, in 1999, for the USA (including a comparison with Canada). In the year 2000, Osberg and Sharpe published international statistics for six OECD countries. This study has quickly become a global reference point and was taken up in the 2001 OECD report on human and social capital. It is especially painstaking in its methodology, while at the same time being as transparent as possible.

One of its original features lies in the fact that it combines the two major types of methods of aggregation and synthesis we have encountered up to this point: the weighted average of heterogeneous variables (used for the indicators described in Chapters II and III) and the monetization of certain variables (indicators in Chapters IV and V). This is the reason why we have left it until now to examine this index.

Another original feature is the inclusion, as one of the four main components of economic well-being, of the degree of economic security (or social protection), which is represented by a particularly stimulating sub-indicator.

Furthermore, of all the major indicators that have acquired an international reputation, this is the one that is probably most likely to provide the basis for a dialogue with the community of national accounting experts, to which its authors belong.

On the other hand, it has the disadvantage of giving a fairly low weighting to environmental variables, which limits its attractiveness for the networks set up by the most militant advocates of sustainable development. However, this disadvantage is surmountable, because the index is explicitly described as being open to public debate with regard to the weighting of the variables, and it does indeed lend itself remarkably well to such debate. It could therefore, be 'greened', which would, in our view, be desirable. Environmental sustainability could be adopted as a major dimension in its own right and, once the matter had been debated, it could be allocated a weighting at least equal to that of the four main dimensions currently in use. The indicator reflects an approach to well-being that ties in well with the social democratic philosophy of the welfare state as a means of reducing economic and social inequalities and risks and investing in the future in these two areas (social Keynesianism). However, increasing awareness of the considerable damage being caused to the natural environment, which is seriously compromising the prospects for the well-being of future generations, should bring about a shift in this approach. This is probably a condition that will have to be fulfilled if this indicator is to achieve even greater international success.

1 Sustainable Well-Being: definition and the corresponding method

According to Osberg and Sharpe (1998, 2003), the four dimensions of economic well-being that would constitute an ideal indicator of economic well-being are:

- effective per capita consumption flows: consumption of marketed goods and services, effective per capita flows of household production, leisure and other non-market goods and services;
- net accumulation of stocks of productive resources: net accumulation of tangible capital, housing stock and consumer durables; net accumulation of human and social capital and investment in R&D; net changes in the value of stocks of natural resources; environmental costs and net change in level of foreign indebtedness;
- income distribution, poverty and inequality: intensity of poverty (incidence and depth) and inequality of income;

- degree of economic security or insecurity: economic security from job loss and unemployment, illness, family break-up, poverty in old age.

These four dimensions form the basis for constructing the indicator. In practice, the ideal list of relevant variables is handled pragmatically, depending on the availability of sufficiently reliable data. Table 7 lists the variables that make up the indicator, grouped together under the relevant dimension of economic well-being, which are all weighted equally. However, these weightings are regarded as open to public and political debate, the only legitimate process for 'revealing collective preferences'. Within each dimension, the component variables (15 in all for all four dimensions) are dealt with in two different ways. For the first two dimensions, which are the most economic in nature, the monetization method is adopted, including for the environmental degradation variable. For the last two dimensions, which are the most social in nature, a weighted average is calculated (see Box VI.1). This dual method seems to us well suited to the dual nature of the variables, and although it seems to complicate things compared with the methods we have encountered hitherto, it may well be an advantage in terms of transparency and openness to debate. In any case, if one thinks about it, an indicator like the HDI (Chapter II) also combines these two methods (monetization and weighted average), since the first of its three component dimensions is nothing other than per capita GDP.

For some of the 15 indicators used, it is clear that the available evaluations are somewhat crude, the worst offender being undoubtedly the estimated per capita value of natural resources. However, it should not be forgotten that the aim here is to construct an index, that is to highlight variations or trends starting from a baseline year; as a result, the influence of some of the (probably enormous) biases in the evaluation of natural resources (or human capital and domestic work, two other tricky problems) may be attenuated when the focus is on variations, that is *relative* values. This attenuation is all the stronger since the evaluations are conducted over long periods of time.

Although the overall principles are the same, it is worth pointing out that the statistical methodology used to construct a comparable

Table 7 The Index of Economic Well-Being (IEWB): components, indicators and methods

Dimensions and value of each variable in Canada (1997 and 1971)	Variables adopted and indications as to method
	Adjustments to marketed personal consumption flows
13 501 (in 1992 Canadian dollars) Before adjustment: 15 548 'Regrettables': 1 839	marketed consumption per capita* (expressed in national currency in constant prices) plus index of changes in life expectancy and adjusted to take account of variations in annual working time per person**.
5 390	government final consumption expenditures, excluding debt service charges (expenditure regarded as 'defensive' is not subtracted here)
7 299	Household work and volunteer work. Estimate of the value of one hour's domestic work on the basis of the hourly rate paid to a replacement worker.
	Per capita stocks of wealth
41 795 (in 1992 Canadian dollars) In 1971: 23 502	Net physical capital stock per capita (monetary valuation): perpetual inventory method where investment flows are accumulated over time, with depreciation rates applied to the different assets.
1 856 In 1971: 788	Research & development capital stock (monetary valuation), valued at cost of investment and assuming a depreciation rate of 20%.
9 159 In 1971: 15 170	Value of natural resource stocks (monetary valuation). Depending on the existing national or international data, attempts might be made to include estimated values for principal mineral resources, forests and energy reserves (which the authors do in the case of Canada).
73 964 In 1971: 52 654	Stocks of human capital: education costs for the entire population, estimated on the basis of cost per year of education expenditures and of estimates of the distribution of education attainment in the population.
-107 573 In 1971: -5 512	Minus: net foreign indebtedness

Table 7 The Index of Economic Well-Being (IEWB): components, indicators and methods – *continued*

400 In 1971: – 362	Minus: cost of polluting emissions (limited to estimates of social costs of CO_2)
Dimensions and value of each variable in Canada (1997 and 1971)	Variables adopted and indications as to method
	Equality
0.952 (base 1 in 1971)	Sen–Shorrocks–Thon measure of poverty intensity (base 1 at beginning of period) constitutes three-quarters of the equality index. It combines the usual poverty rate, the average poverty gap ratio and an inequality of poverty gap ratio
0.971 (base 1 in 1971)	Gini coefficient (income after tax), constitutes one quarter of the equality index
	Economic security (see section 2 below)
	Economic risk associated with unemployment
	Economic insecurities associated with illness
	Economic risk associated with single-parent poverty
	Economic insecurity in old age

* In unit, per unit of consumption, in the sense of household consumption statistics.
**This method makes it possible to take account of free time gained (or lost) without embarking on the always problematic task of estimating the monetary value of total free time (as Nordhaus & Tobin do). It amounts to assuming that the value individuals attribute at the margin to free time gained (excluding unemployment) is equivalent to their net wage after tax.

indicator for several countries tends to differ in certain respects from that used for a single country, because of differences in statistical sources, available surveys and so on. For example, the value of domestic and voluntary work, which was included in the calculation of Canada's economic well-being, could not be included in the subsequent international comparison for lack of reliable data. Thus a clear distinction has to be made between what an ideal indicator would be and what it is possible to measure at a given time. The fact that reliable data on an essential element of well-being are not available for a particular country is often a major political question and not a technical one (see our comments on domestic work in Chapter IV).

2 The measurement of economic security in the IEWB

This fourth dimension of the IEWB is certainly the most novel, by virtue both of the solidity and universality of its political and moral reference points and of its method. This is why we intend to examine it more closely here, sometimes citing whole sentences from Osberg & Sharpe (2003). *Economic* insecurity is defined as 'the anxiety produced by (...) an inability to obtain protection against subjectively significant potential economic losses' Individuals' perception of insecurity is the result of their expectations for the future, combined with their current economic situation, which explains why it cannot be properly represented by measures such as the ex-post variability of income flows.

Rather than attempting to devise an overall measure of economic insecurity, the authors adopt a 'named risks' approach based on four key economic risks, legitimated by reference to the United Nations' Universal Declaration of Human Rights:

> Everyone has the right to a standard of living adequate for the health and well-being of himself and of his family, including food, clothing, housing and medical care and necessary social services, and the right to security in the event of unemployment, sickness, disability, widowhood, old age or other loss of livelihood in circumstances beyond his control.
>
> (Article 25)

Taking this international reference point as a starting-point, the authors of the IEWB seek to measure variations in the 'objective' economic risks associated with unemployment, sickness, widowhood (or single parenthood) and old age. In each case, the risk of economic loss linked to the event in question is evaluated as 'a conditional probability, which can itself be presented as the product of a number of underlying probabilities'. The prevalence of the underlying risk is weighted by the proportion of the population that it affects. The core hypothesis is that changes in the subjective level of

---

Box VI.1   Methods of evaluating economic security (summary of Osberg & Sharpe 2000)

Unemployment

The economic risk associated with unemployment can be modelled as the product of the risk of unemployment in the population and the extent to which people are protected from the income risks of unemployment. The extent to which people have been protected by unemployment insurance from the financial impacts of unemployment can be modelled as the product of (a) the percentage of the unemployed who claim regular unemployment insurance benefits and (b) the percentage of average weekly wages replaced by unemployment insurance (the gross replacement rate).

Illness

The indicator focuses on the risk of large out-of-pocket healthcare costs, with the risk directly proportional to the share of private medical care expenses in disposable income

Single-parent poverty

When the UN Universal Declaration of Human Rights was drafted in 1948, the percentage of single-parent families was relatively high in many countries, partly as a result of World War II. At that point of time, 'widowhood' was the primary way in which women and children lost access to male earnings. Since then, divorce and separation have become the primary origins of

(continued)

Box VI.1   *continued*

single-parent families. However, it remains (statistically) true that many women and children are 'one man away from poverty', since the prevalence of poverty among single-parent families is extremely high. As the authors put it:

> To model trends in this aspect of economic insecurity, we multiply the probability of divorce by the poverty rate among single female parent families and by the average poverty gap ratio among single female parent families. The product of these last two variables is proportional to the intensity of poverty. ... It should be emphasized that in constructing a measure of the economic insecurity associated with single parent status, we are not constructing a measure of social costs of divorce. Economic well-being is only part of social well-being, and divorce has emotional and social costs that are not considered here.

Old age poverty

The idea of insecurity in old age is modelled as a chance that an elderly person will be poor, and the average depth of that poverty. Thus the method is the same as in the previous case.

Overall index of economic security

The four risks described above are aggregated into an index of economic security. The aggregation weights are the relative importance of the four groups in the population, namely:

- for unemployment, the proportion of the population aged 15–64 in the total population;
- for illness, the proportion of the population at risk of illness, which is 100 per cent;
- for single-parent poverty, the proportion of the population comprised of married women with children under 18;
- for old-age poverty, the proportion of the population in immediate risk of poverty in old age, defined as the proportion of the population aged 45–64 in the total population.

anxiety about a lack of economic safety (variations in subjective well-being) are proportionate to changes in objective risk.

This indicator, it should be noted, is very different from the personal security indicator described in Chapter III, which combines measures of objective risks (the unemployment rate, for example, although there is no attempt to impute a monetary value to the associated losses of resources) and measures of feelings of insecurity (opinion polls). It also takes account of physical insecurity and considers poverty as a risk.

3 The main results

Figures 8 to 12, which were sent to us by the authors and are published here with their permission, illustrate the significance of their research.

Figures 8, 9 and 10 relate to Canada during the period 1971 to 1997. Figure 8 shows the evolution of the IEWB and its major components. It is clear from this that the main factors dragging the index downwards during a period of growth in material consumption and stocks of wealth are the inequality and, above all, the economic-insecurity indicators. Figure 9 compares the trends in GDP and the IEWB in Canada. Figure 10 compares the evolution of the IEWB, the Index of Social Health (calculated for Canada by Brink and Zeesman 1997), the GPI (see Chapter V) and Nordhaus & Tobin's SMEW (Chapter IV). These two indices were calculated for Canada by Messinger and Tarasofsky (1997). Except for the ISH, there is considerable similarity in the profiles of these curves.

Figures 11 and 12 are taken from Osberg & Sharpe (2002). They relate to the UK and Norway during the period 1980–99. The profiles are sharply contrasting. On the one hand, the UK, whose GDP increased by 50 per cent over the period, saw its IEWB collapse between 1985 and 1991, so much so that, despite a slight rise since that date, it was still very considerably below its 1980 level at the end of the period. It was at the end of 1990, of course, that Margaret Thatcher's three successive periods of office came to an end after more than 11 years. In Norway, on the other hand, there was simul-

Figure 8   Canada: the IEWB and its components, 1971–99 (1971 = 1)
(*Source*: provided by Osberg and Sharpe)

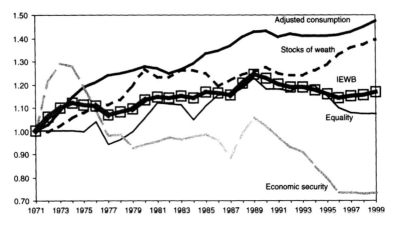

Figure 9   Canada: GDP per capita and IEWB, 1971–99 (1971 = 1)
(*Source*: provided by Osberg and Sharpe)

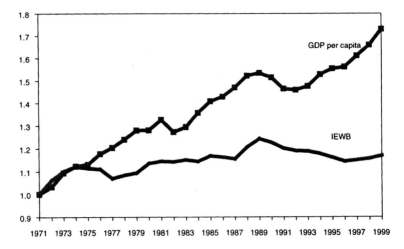

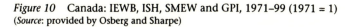

Figure 10   Canada: IEWB, ISH, SMEW and GPI, 1971–99 (1971 = 1)
(*Source*: provided by Osberg and Sharpe)

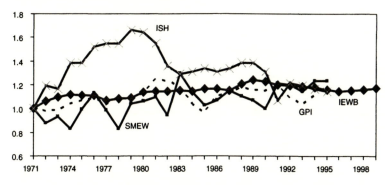

taneous growth in GDP (slightly more than in the UK) and the
IEWB, although the rates of growth were different. However, it is
fairly clear that there is no reason why these rates of growth should
be similar, either in this case or in that of the Index of Social Health
(see Box III.4 entitled 'The need for caution...'). The most significant
thing about these graphs is the opportunity they provide for
judging whether or not overall economic progress has been accompanied by a 'minimum' level of progress in the IEWB and to debate
that 'minimum' level.

Figure 11   GDP per capita and IEWB in the UK 1980–99
(*Source*: Osberg & Sharpe 2002)

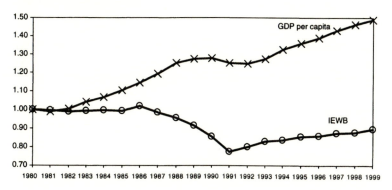

Figure 12   GDP per capita and IEWB in Norway, 1980–99
(*Source*: Osberg & Sharpe 2002)

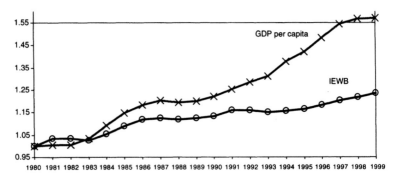

VII
Evaluating the Indicators

Evaluating the indicators outlined in this book involves assessing their effectiveness and their ability to produce meaning as tools for measuring 'societal performance' and, beyond that, as cognitive frameworks for public or private action. However, there is no clear, universally accepted definition of what constitutes societal performance, either in the terminology used – expressions such as '(economic) well-being, 'development', 'social progress' and 'quality of life' are deployed in turn – or in the conceptualization of that performance.

This being the case, any attempt to evaluate the indicators necessarily brings into play a wide range of criteria, which may also be used in combination with each other. Drawing on a number of previous studies (Sharpe 1999; Perret 2002), we aim in this chapter to develop a multi-criteria framework for evaluating the major synthetic indicators that can be regarded as an aid to deliberation or decision-making.

Some of the criteria used in this framework can be directly positioned on a sliding scale ranging from ++ to – –. This is the case, for example, with methodological transparency: an indicator has a more or less transparent methodology, which the actors can therefore appropriate more or less easily. In the case of another, larger group of criteria, the quantitative or normative positioning is more ambivalent, since it requires the adoption of a theoretical stance or one based on judgements or social conventions, or it is quite simply not conceivable. This is the case, for example, with the 'number of dimensions' criterion: where does the optimum choice of number lie? Here, compromises have to be made between the objective of reducing the complexity of the indicator (by adopting only a small number of dimensions, for example) and that of producing as comprehensive a measure as possible of social progress. However, the whole exercise is made all the trickier by the fact that some synthetic indicators are both complex and not very comprehensive. This applies to GDP, for example.

Thus such choices need to be debated and also require a certain degree of pragmatism. No purely technical explanation can in itself give legitimacy to any particular solution.

1 The (relatively) technical aspects of the indicator

We noted in Chapter IV that one of the conditions for the emergence of a consensus around an alternative convention for measuring wealth was technical feasibility and the narrowing of the margins of uncertainty for both the data and the results. With this in mind, we will outline various criteria for differentiating and evaluating the indicators. These criteria relate to: (a) the method of construction (we will not review in any detail the methods of aggregation, since they have already been the focus of some attention in other chapters) and the criteria for selecting and weighting the various dimensions and component variables; and (b) their technical potential for use. Over and above their apparent technical suitability, these questions also bring into play the political and ethical conventions that frequently underlie the major technical choices forming the background to the expert debates (see next chapter).

The method of construction

Five criteria can be used here to differentiate the indicators from each other:

1. The nature of the initiative: the aim here is to ascertain whether the various alternatives selected (dimensions, components, weightings) were the product of individual or collective initiatives (research institutes, networks, etc.) and who funded the projects. It might reasonably be assumed that an indicator is more likely to be of lasting value when public institutions have contributed to its construction.

2. Accounting consistency: this applies mainly to monetised indicators. The priority here is to evaluate the robustness of the conventions adopted for the monetization of certain variables. We will return to this point in Chapter VIII, where we note in particular that the adoption of 'debatable' conventions does not necessarily imply a 'reduction in quality', particularly if those conventions have been the subject of real debate.

3. The 'integrity' of the data on which the indicator is based: that is the degree of reliability in the choices of variables, data and sources.

4. Openness to the possibility of taking account of missing data or of 'invisible' or 'forgotten' populations. Many initiatives take as their starting-point 'what exists' in terms of databases. Such pragmatism is very understandable, but it can lead to an extremely restrictive choice of variables. Conversely, some of the initiatives mentioned here have led to the development of an ideal conceptual framework that provides fairly comprehensive coverage of the social questions under the spotlight. 'For lack of available data', these initiatives subsequently resort to measures based on existing data, while at the same time highlighting the importance of collecting new data. These approaches have the advantage of providing some pointers towards more comprehensive information systems (this is the approach advocated by Osberg and Sharpe for their IEWB, for example; see Chapter VI).

5. The conceptual framework: are the concepts the indicator is seeking to evaluate clearly defined and are they integrated into a coherent, convention-based framework that has sufficiently broad support?

The choices of dimensions and variables

We have adopted four criteria:

1. The completeness of the dimensions and variables: do the dimensions selected encompass in a reasonable and satisfactory way the whole of the reference concept, and in all of its individual dimensions? does the indicator being evaluated incorporate the economic, social and environmental dimensions?

2. Is the indicator being measured 'objectively' and/or subjectively?

3. The number of dimensions: is there a 'balance' between the various dimensions or is one of them over-represented (Sharpe 1999, p. 13)? what convention-based criteria were used in selecting the dimensions?

4. The choice and transparency of the weightings used for the various dimensions and variables: what proportion of the construction methods seek some form of legitimacy, whether it be obtained through so-called 'expert' groups (BIP 40), opinion surveys or the individuals' free choice (the Dashboard, for example)? There should also be some investigation of the degree of transparency or opacity in the choice of weightings.

The indicator's potential for possible alternative uses

Five criteria will be used:

1. The continuity of the data over time. This is sometimes an element in the method used to construct a composite indicator (e.g. when the values of the different variables are normalised by using the 'best' and 'worst' of the years observed). More fundamentally, however, most of the judgements of progress are meaningless unless they are located in time.

2. The international comparability of the variables and sources.

3. The universality of the indicator. The question of universality can be posed at two levels, which are not directly connected with each other.

    • The first is international universality, which has the advantage of immediate comparability but also brings with it the risk of excessive institutional influence. In this regard, we do not see it necessarily as a failing if an indicator is country-specific. This rootedness in a particular national institutional setting has implications for international comparability, but may well prove to be a much more effective way of measuring a country's social progress, particularly if that very rootedness gives the indicator a certain degree of legitimacy.

    • The question of universality must also be posed at the national level. If a synthetic indicator is to have both credibility and legitimacy, it is essential that its reference points are not constantly changing over time. However, this is a tricky issue (since the status quo may also have disadvantages), to which we will return in the final chapter with regard to the relation-ship between politics and national accounts.

4. Can the indicator be broken down into signifying dimensions? Such disaggregation can be undertaken in two different but complementary ways. Firstly, it may be advantageous to be able to give the indicator different values, depending on category (gender, age group, region etc.), as is the case with some of the initiatives described here. Secondly, and most impor-tantly, if an indicator can be disaggregated, it becomes a suit-able basis for analyses of individual dimensions (such as poverty, inequalities, economic insecurity, environment, etc.). In this way, the strengths and weaknesses of a society's progress can be readily identified and public policies or private strategies appropriate to each dimension can be developed.

2 Evaluating the indicators as policy-making aids

We have adopted four criteria as a basis for evaluating the effective-ness of the indicators as aids in public decision-making:

1. Is there any plan to make use of the final result, and if so how is it to be used? If the aim is simply to monitor an indicator of societal wealth over time, then the use to be made of it is implicit: if the indicator goes up or down, it is easy to infer that the society in question is either progressing or declining. However, in the case of indicators developed in order to com-pare countries, for example, what should be considered a good performance? This type of use often requires the results to be 'normalized' by adopting reference points regarded either as very good or very bad (e.g.: the best and worst results over a reference period or for a group of countries).

2. Does the indicator have a clearly defined economic and/or social-policy objective? This question is linked to whether or not it is possible to break the synthetic indicator down into dif-ferent dimensions. In fact, most of the synthetic indicators examined in this book can be meaningfully disaggregated.

3. The possible uses of the indicators: these are diverse and range from monitoring to the circulation of a public report, via fore-casts or material for public debate, although no normative

judgement can be made as to the relative value of these various uses.

4. The legitimacy the indicator has acquired, its media profile and influence in public debate. This may be judged in part on the basis of the indicator's lifespan. Our current knowledge is somewhat limited in this regard, because most of the alternative indicators have emerged only recently and many of them are still in the process of acquiring legitimacy.

The multi-criteria evaluation of synthetic indicators of economic well-being or national wealth is just one stage in formulating a judgement of such indicators. In order to be effective, these indicators must be relocated within frameworks that fulfil the threefold objective of expertise, pluralism and democracy.

None of the synthetic indicators examined here can lay claim to 'full marks' for all these criteria, and in any case the scores that can be envisaged are themselves a function of the possible uses.

By way of illustration, the following tables offer a multi-criteria analysis of some of the indicators examined in the previous chapters. GDP and the HDI were selected because of their media profile and their influence on cognitive representations of a certain form of progress, together with the IEWB, which seems to be a promising indicator.

Table 8 Evaluation of indicators from a policy perspective

	GDP	HDI	IEWB
Clear economic and/or social policy objective	++ Objective: raising the rate of economic growth	+ for the three dimensions of the HDI	+ presented as a supplement to GDP
Direct public decision-making tool	+ Specific economic policy tool	+ International report and advocacy tool	+ Supplementary tool for GDP
Acquired legitimacy, media profile, influence on public debate, lifespan of indicator	Established legitimacy, high media profile, influential in public debate, long lifespan.	++ Rapid diffusion through international organization and the legitimacy of its designers.	++ Increasing influence, particularly as a result of its authors' efforts to publicize it internationally.

Notes: ++ good indicator
+ adequate indicator
– less man adequate indicator
– – bad indicator

Table 9 Technical evaluation of the indicators: three examples

	GDP	HDI	IEWB
Individual or collective method of construction	Collective, but internal debates confined to expert circles. Publicly funded.	Collective: UNDP, international debates Funding: international organization	Two researchers Funding: private foundations. Some Canadian public initiatives.
Accounting consistency	+/– consistent; uncertain conventions (services)	Non-monetary mode of valuation	Essentially non-monetary mode of valuation
Integrity of data	+/– yes; unreliable data for non-market activities	++	+
Possible openness	+ The definition of wealth can be extended	– Not explicitly open. Supplementary indicators produced by the UNDP.	+ High level of potential openness.
Conceptual framework	Market consumption is the only aspect of wealth taken into account.	Sen's *capabilities* theory	+ Fairly coherent framework, precise definition of economic well-being
Completeness relative to a global concept of well-being	– – GDP is not a measure of well-being	Only 3 dimensions. Supplemented by the HPI and GEM.	– Focuses on *economic* well-being. Little attention paid to the environment.
Objective or subjective measure	Objective measure	Objective measure	Objective measure
Number of dimensions	High degree of complexity	3 dimensions High degree of simplicity	4 dimensions, 15 variables

Table 9 Technical evaluation of the indicators: three examples – *continued*

	GDP	HDI	IEWB
Choice and transparency of weighting	Monetized indicator; transparent only to a small circle of experts	Simple average	+ Simple average of the 4 dimensions. Transparent. Possible choice of weightings.
Continuity over time; consistency	++	++	++
International comparability, universality vs. national rootedness	Dual rootedness	Dual rootedness	National rootedness, but international comparisons
Manipulability–disaggregation	++	++	++ for each of the four dimensions of economic well-being

Notes: ++ good indicator
+ adequate indicator
– less than adequate indicator
– bad indicator

VIII
Sustainable Human Development and National Accounts in the Twenty-First Century

Our purpose in this chapter is to outline some ideas about the shape of national accounts in the twenty-first century. To that end, it is essential briefly to review the history of national accounting and the current controversies in the field.

1 Two perspectives on the history of national accounting

Our argument takes as its starting-point a simultaneous reading of two of the foremost French books on the history of national accounting, published twenty years apart. The first is that by François Fourquet (1980), the second the monumental survey by André Vanoli (2002), one of the leading specialists in the world in national accounting since the 1960s.

Readers of Vanoli's study will gain the clear impression that the history of national accounting since the 1930 and 1940s is one of continuous technical progress. From this perspective, the methods and concepts that have emerged out of international debates among experts have led to continuous improvements in a theoretical and practical tool that is capable of representing, with increasing relevance, actual macro- and meso-economic relations. It is the history of a major discovery, made by scholars in the course of a process characterized by academic disputes and debates among the various schools that have gradually emerged. As a narrative, it is not dissimilar to the story of the quest to find a proof for Fermat's Theorem told with considerable verve by Simon Singh (2002). The following are the key words in this story: *rigour* (the permanent academic objective), *difficulties, ambiguities, gaps, defects* and *shortcomings* (which can be overcome by logical academic debate), *grey areas,* which become clearer, and, finally, *harmonization, consistency* and *integration,* which denote the most accomplished construct (which continues to be improved), namely the integrated systems of national accounts of 1993 and 1995 (SNA 93 and ESA 95). Vanoli sketches in the historical and political context (slump, Keynesian macro-economics, increasing role of the state), particularly in the last chapter of his book. However, politics do not emerge as having played a decisive role either in the debates, which are largely the province of leading academic experts, or in the solutions finally adopted, which are the product of an intellectual process of 'harmonization'. True, the disputes are never-ending and are still ongoing; however, they are essentially disputes among academics and the various schools of thought that have grown up, which ultimately add to the sum of knowledge in the area and thereby produce a 'common good'.

The history of national accounts as told by François Fourquet in the case of France is based on the (recorded) narratives of 26 major actors. It is a completely different story from the one Vanoli tells, since the construction of these statistical tools and the accompanying controversies are seen here as determined largely by politics and political views of national power and wealth. The history of the French system of national accounts is not that of an 'intellectual genesis' but rather of a 'political genealogy' (p. 137); from this perspective, national accounts are primarily accounts of 'national power' as conceived by politicians in France (as well as in other countries) in the course of this period.

Let us take one major example. Until 1976, the activities of central and local government (a sector subsequently given the appellation 'non-market services') were not part of national output in the French system. This was no academic 'mistake' or 'oversight', to be rectified subsequently. Rather, it reflected a political desire to reconstruct the country by putting in place an ambitious industrial policy and giving priority to competitive market activities which, in the political thinking of the time, were the mainstays of French 'power'. The subsequent convergence of the French and 'Anglo-Saxon' concepts and systems cannot be explained by any progress in 'mutual understanding' among experts (to use Vanoli's words, quoted by

Fourquet) but mainly by the politically driven convergence of the conventions by which economic policy priorities are determined. The much vaunted theoretical and technical 'harmonization' was subordinated to this political convergence.

Academic debate and the socio-political conventions

In our view, Fourquet's historical interpretation goes a long way towards explaining the way in which the successive systems of national accounts were 'discovered' under the influence of the major political choices made during each period. However, the almost complete absence in Fourquet's narrative of any reference to the debates among specialists, which is the counterpart of Vanoli's almost total disregard of the influence of politics, does pose various problems. Firstly, the academic debates take place in relative autonomy and require sequences of logical arguments and 'statistical thinking', some of which have nothing to do with the political conventions on wealth and power. Secondly, politicians and policy-makers cannot wholly disregard the debates among recognized experts (who themselves come from a range of different political backgrounds and often defend national points of view in international circles) because, in countries other than those governed by authoritarian regimes, the major players in politics have publicly to justify the choices they make. The possibility of opposition from experts with their own convictions and a willingness 'to put their heads above the parapet' is one of the risks to be guarded against. Thirdly and finally, once the statistical frameworks have become established and been socially validated, they become as much constraints as resources for politicians and policy-makers; in this sense, they are rules, similar to those enshrined in law, from which it is not easy to break free.

Despite what we would regard as its excessive political determinism, we are inclined to take seriously the argument that the 'major' political conventions exert considerable influence over the construction of the 'major' statistical tools used in economic policy-making. It has to be said that Fourquet and his 26 witnesses provide many examples that support that argument. Moreover, we have other, more recent examples that point in the same direction (see §2 and Box VIII.1).

In order to avoid any misunderstanding, we should make it clear that this argument concerning the highly political nature of the major choices made during the development of systems of national accounts is not in any way intended to imply any dishonesty or politicking on the part of statisticians 'obeying orders'. Rather, the use of the term 'political' in this argument is intended to denote a somewhat 'loftier' vision of the major national and international choices that generally underlie shifts of political direction over the medium term. For their part, the experts in question carry out their work rigorously, innovate, initiate debates and make improvements to the statistical tools used to capture economic practices. Like everybody else, however, these experts are steeped in the existing cognitive frameworks (particularly when it comes to notions about wealth and the 'right' economic policies). At the same time, they are part of national or international institutions that have their own political logic and are also dependent on the political sphere. Finally, they have an obligation to serve the economic policies currently in force and the states (or international bodies) that employ them, while at the same time defending – not unsuccessfully as it happens – their intellectual independence and sense of professional ethics.

Accounts of national power... embedded in socio-political conventions

The continuous developments and progress made in the field of national accounting are neither the fruit of debates within an intellectual discipline supposedly unaffected by the tensions of the social world nor a passive reflection of wider political ideas. Rather, they resemble a pursuit or chase between 'management tools' (which are constantly being improved) and an 'enterprise' (a national one in this case) whose organization, strategy and values (i.e. political conventions) change over time. The development and improvement of these management tools brings into play national 'power' (i.e. the exercise of political power), the conventions that give expression to the strategy and objectives adopted and logical arguments propounded within the statistical and accounting communities. François Fourquet is quite right to emphasize the role of 'power' (the power of a nation or state, in this case), but what influences the debates on national accounts is the pairing of power and conventions on wealth: the second element in this pairing denotes, as it were, the cognitive and symbolic content of the exercise of power, the meaning of the arguments advanced in its justification.

Box VIII.1   On the conventions underlying national accounts

Experts in national accounts know better than anyone that their work is based on conventions, including those used to calculate GDP: convention-based nomenclatures, statistical conventions on methods of gathering and processing statistics and change in product quality, conventions for evaluating the output of banking, insurance, health and accommodation services, retailing and so on.

However, these *statistical conventions* – which arise out of a need to make choices between competing methods and procedures that are, on the face of it, equally valid, with an element of arbitrariness entering into the selection – are different in kind from the conventions on wealth and well-being that interest us here. These latter concern the overall representation of what counts and what should be counted when a nation's wealth is evaluated and the contribution to well-being made by various activities and assets. The first and second sets of conventions (statistical conventions, on the one hand, and socio-political conventions on what constitutes wealth, on the other) are not, of course, unconnected. It is clear, for example, that the conventions used to evaluate health services in terms of their impact on a nation's health and the increase in life expectancy do not reflect the same vision of wealth as those conventions that measure the volume of medical interventions. However, these two categories of conventions (one statistical, the other related to wealth) are also relatively independent of each other. And it is the second category that is giving rise to more questions outside the circle of expert statisticians. It is these conventions whose legitimacy is disputed and whose diversification is proposed in the initiatives outlined in this book. The first category consists of socio-technical conventions which, to varying degrees, incorporate non-technical considerations of 'what really counts'. The second consists of non-technical conventions that influence the (statistical) technique but are located well upstream. They are expressed in terms of value judgements, analysis of which falls more within the scope of moral and political philosophy (or a form of political economy intent on rediscovering its moral origins) than of standard economic and statistical expertise.

2 The current controversies: as political and moral in nature as the early debates

The best examples of recent controversies in the field of national accounting are to be found in André Vanoli's book (2002). They are lucidly explained in terms of the range of options that present themselves. The main actors involved are leading international specialists, and the author frequently alludes to his own preference for the solutions that are, in his view, the most 'coherent' and 'rigorous'. In most cases, however, there are good grounds for thinking that these antitheses between apparently technical and purely theoretical solutions actually conceal 'political preferences' (at national or international level) and normative judgements that have nothing technical about them and are linked to both the exercise of 'power' and the prevailing 'socio-political conventions'.

This is illustrated particularly clearly by the controversial case of 'military capital goods', particularly 'destructive armaments' (p. 390 ff.), which can be treated either as productive investments (GFCF) or as current consumption, with obvious effects on the structure of GDP. How can one not be struck by the fact that American national accountants are in the forefront of those advocating that armaments should be treated as investments, their 'cover' of course being a technically 'neutral' argument, namely that 'armaments should be regarded technically as providing a service that contributes to national defence'? This is an argument to which Vanoli objects, on the grounds that 'in the event of war, military operations, which are self-evidently destructive, cannot be regarded as a production process'. We will not seek to come to any verdict on this controversy, but it clearly falls within the scope of political ethics. In the 1950s, incidentally, the founders of the French system of national accounts saw armaments as the 'symbol of evil' and the worst possible example of an 'unproductive and sterile burden' (Fourquet 1980, p. 149).

Are armaments an isolated example, the only one in which an accounting controversy visibly introduces a moral and political dimension into an area otherwise wholly dominated by academic and technical considerations? This is absolutely not the case, as examination of some of the many controversies around 'imputation' and 'redirection' shows. We will confine ourselves to the latter and look at one major example. The 'redirection' of flows of social security contributions involves not recording the transactions where they take place (deductions from employees, on the one hand, and from firms, on the other) but putting them together under a different heading. The logic deployed here is an explicitly economic one. For example, all social security contributions related to employees, regardless of who actually pays the sums in question, might be grouped together under the same heading in order more clearly to represent these contributions in their entirety as an 'indirect wage'. Is there anyone who cannot see the purely political issue at stake in an accounting decision of this kind, and what its abandonment might mean in terms of perceptions of the foundations of social protection?

It would be possible to take virtually all the controversies examined by Vanoli (intangible investment, capital gains and losses, the environment, etc.) and interpret them in the same way. Clearly, arguments about technical consistency and epistemological cogency play a role, but their function is, in part, to represent and translate much more fundamental political conventions relating to wealth and what makes a healthy economy and a good society into the discourse of experts, who are, quite legitimately, concerned to ensure a certain degree of coherence and consistency. This has already been clearly demonstrated, in Chapter IV, by the very important example of the treatment of activities carried out within the home.

Conclusion

The hypothesis that social and environmental considerations will increasingly gain ground and lead to the development of 'accounts for the twenty-first century' based on the notion of sustainable human development (or on similar notions, such as sustainable well-being, for example) comes up against one powerful objection. A similar rise to prominence has already happened once before, in the 1970s. It collapsed like a soufflé from the end of the decade onwards when, against a background of recession, inflation and unemployment, the political order (both nationally and internationally) was more or less as follows: 'let's go back to our core business: growth and competitiveness'.

There was unanimous agreement among the actors of the period as to the reality of this sudden about-turn. Incidentally, André Vanoli describes this about-turn in the following terms, and in doing so provides further proof of the influence of politics on economic and social statistics (quoted in Fourquet 1980, p. 359: these statements date from the early 1980s):

> Several years ago, when we were still in the period of healthy growth, everybody swore by the environment and well-being. People were heard to say: "national accounts are no longer of any value and no longer serve any useful purpose; on the contrary, they merely give rise to confusion". And then, as unemployment and inflation rose, there was a return to absolutely standard concerns. The fashion has changed. A few years ago, we had to do everything, and do it immediately; now, we have to sweat

blood in order to continue to do anything at all to improve the measurement of well-being. What is frequently annoying for the technical experts who have to produce data is the fickleness and impatience their political and administrative masters display as soon as these new attitudes emerge.

This is remarkable evidence of the scale of this political about-turn, its influence on the production of statistics and the way in which the technical experts experience these reversals – not without some justification – as a manifestation of a fairly irresponsible fickleness. However, in the case of the 1980 about-turn, there is every indication that it had absolutely nothing to do with political fickleness but was rather a manifestation of a groundswell whose effects are still being felt today and which started with the 'conservative revolution' of the Reagan and Thatcher era. The basis of a new financial capitalism, of which stock markets and equity ownership are the key components, was then politically constructed by national governments and most of the international institutions.

Be that as it may, this turnabout leads us to wonder whether the rapid rise in the prominence, over the course of the 1990s, of initiatives that sought to extend and bring up to date studies that were begun in the 1970s is not part of a cyclical fashion that the increase in unemployment since 2001 will eventually crush in its turn. Ideas also have their fashions that come and go, as Hirschman (1982) showed. Thus we cannot entirely rule out the possibility that a worsening of the global economic and social crisis might undermine the position of those who advocate that the religion of growth should be put into perspective. However, this seems to us unlikely. Firstly, the global political and ideological context at the beginning of the twenty-first century is not the same as it was in 1980. There is little risk, for example, that global awareness of the gravity of environmental questions will decline. And it can reasonably be assumed that the countries of the South will make their voice heard more loudly as it is relayed to the North via networks, NGOs and some national governments. However, the main difference between the present day and the 1970s lies in two phenomena related not to global geo-political considerations but rather to the renewal of political practice. The first of these phenomena is the emergence of local or regional initiatives and of militant networks. The alternative indicators that emerged during the 1970s were mainly the work of fairly isolated researchers or high-level institutions. Those of the 1990s were established largely by NGOs, associations, foundations and networks, as well as by local and regional 'communities'. In this last case, the need to evaluate the quality of life, of social relations and of the environment is growing apace. We have not analysed these phenomena, but it is quite clear that these preoccupations have been 'decentralized' and are being confronted within networks.

The second phenomenon is the growing influence of women in political life, which is helping to raise the profile of the human, social and environmental criteria of development. Gendered representations of wealth, development and progress certainly exist. The sociologist Françoise Héritier (2002) provides a spectacular example of this:

> Recently, a public opinion survey was carried out by sociologists in order to discover what the main events of the 20th century were. A majority of men said it was the conquest of space. For 90 per cent of women it was the right to contraception.

Thus these observations lead us to believe that, in the years to come, the features that will distinguish the present period from the 1970s will be:

1. the existence of 'participative networks' that are not limited to circles of specialists and leading experts but are run directly by sections of civil society and social movements;
2. the increasingly important role of women in politics; and
3. the increasing political influence exerted by the Southern countries on the definition of the conventions by which development is evaluated.

None of these three trends, which are both forms of and conditions for sustainable human development, is irreversible, but conservatives of all political persuasions will find it very difficult to contain them.

Notes

Introduction

1 Indicators of social capital, as popularized by Putnam (1995, 2000) and taken up by the OECD (2001), are not presented in this book. They have attracted strong criticism (Ponthieux 2003) and do not seem able to provide a reliable basis for international comparison.

III Human Development and Social Progress

1 Example: if the same basket of goods representative of French and American consumption costs 0.9 euros in France and 1 dollar in the USA, the purchasing power parity between the two countries is said to be 1 dollar for 0.9 euros and French GDP can be expressed in dollars (and vice versa).
2 See, for example, the following websites: www.sustainablemeasures.com and www.sustainable.org/creating/indicators.html

IV The First Extensions of GDP

1 His actual words were:

> This shortcoming of the theory in confrontation with the new findings, has led to a lively discussion in the field in recent years, and to attempts to expand the national accounting framework to encompass the so-far-hidden but clearly important costs, for example, in education as capital investment, in the shift to urban life, or in the pollution and other negative results of mass production. These efforts will also uncover some so-far-unmeasured positive returns – in the way of greater health and longevity, greater mobility, more leisure, less income inequality, and the like.

> (*Nobel Lectures, Economics 1969–1980* ed. Assar Lindbeck, World Scientific Publishing Co., Singapore, 1992; http://www.nobel.se/economics/laureates/1971/kuznets-lecture.html.)

121

References

Baneth J. 1998, 'Les Indicateurs synthétiques de développement', *Futuribles*, 231, May.

Brink S. & Zeesman A. 1997, *Measuring Social Well-Being: an Index of Social Health for Canada*, Human Resources Development, Canada, June. Research document R-97-9. Online at http://www.hrdc.gc.ca/stratpol/arb/publications/research.

Brown L. 2001, *Eco-Economy: Building an Economy for the Earth*, W.W. Norton, New York.

Cobb C. & Cobb J. 1994, *The Green National Product: a Proposed Index of Sustainable Economic Welfare*, University of America Press, Lanham, MD.

Cobb C. & Daly H. 1989, *For the Common Good. Redirecting the Economy toward Community, the Environment and a Sustainable Future*, Beacon Press, Boston.

Cobb C., Halstead T. & Rowe J. 1995, *The Genuine Progress Indicator: Summary of Data and Methodology*, Redefining Progress, San Francisco.

Cobb C., Glickman M. & Cheslog C. 2001, *The Genuine Progress Indicator Update*, Redefining Progress, Issue Brief, December.

Estes, R.J. 1998, 'Trends in World Social Development, 1970–1995. Development Prospects for a New Century', *Journal of Developing Societies*, 14.

Everett G. & Wilks A. 1999, *The World's Bank Genuine Savings Indicator: a Useful Measure of Sustainability?* Bretton Woods Project, at http://www.brettonwoodsproject.org.

Fourquet F. 1980, *Les Comptes de la puissance*, Encres, Paris.

Gadrey J. 1996, *Services, la productivité en question*, Desclée de Brouwer, Paris.

Gadrey J. 2002, *The misuse of Productivity Concepts in Services*, in Gadrey J., Gallouj F., *Productivity, Innovation and Knowledge in Services*, Edward Elgar Publishing, Northampton.

Gadrey J. 2003, *Les Conventions de richesse au cœur de la comptabilité nationale. Anciennes et nouvelles controverses*, Colloque Conventions et institutions, Université de Paris X-Nanterre, 11–13 December.

Gadrey J. & Jany-Catrice F. 2003, *Les Indicateurs de richesse et de développement. Un bilan international en vue d'une initiative française*, Research Report for DARES, March. Available online at: http://www.travail.gouv.fr/etudes/etudes_g.html.

Hamilton K. 2001, *Measuring Sustainable Development. Genuine Savings*, Round Table of the OECD, 31 May. See: http://www.oecd.org/dataoecd/21/12/2430203.pdf.

Hamilton K. & Clemens M. 1999, 'Genuine Savings Rates in Developing Countries', *World Bank Economic Review*, 13(2): 333–56.

Heritier F. 2002, *Masculin/Féminin II*, Odile Jacob, Paris.

123

Hirsch F. 1976 (2nd edn 1995), *Social Limits to Growth*, Routledge, London.

Hirschman A. 1982, *Shifting Involvements: Private Interest and Public Action*, Princeton University Press, Princeton NJ.

Jackson T. & Stymne S. 1996, *Sustainable Economic Welfare in Sweden. A Pilot Index 1950–1992*, Stockholm Environment Institute, Stockholm. See: http://www.sei.se/pubs/dpubs.html.

Messinger H. & Tarasofsky A. 1997, 'Measuring Sustainable Economic Welfare: Looking Beyond GDP', paper presented at the Annual Meeting of the Canadian Economics Association, St John's, Newfoundland, June 2–4.

Miringoff M. & Miringoff M-L. 1999, *The Social Health of the Nation. How America is Really Doing?*, Oxford University Press, Oxford.

Miringoff M-L., Miringoff M. & Opdycke S. 1996, 'The Growing Gap between Standard Economic Indicators and the Nation's Social Health', *Challenge*, July–August.

Nordhaus W. & Tobin J. 1973, 'Is Growth Obsolete?' in *The Measurement of Economic and Social Performance, Studies in Income and Wealth*, Vol. 38, Columbia University Press for the National Bureau of Economic Research, Clawson.

OECD 2001, *The Well-Being of Nations: The Role of Human and Social Capital*, Paris, OECD.

OECD 2002, *Aggregated Environmental Indices, Review of Aggregation Methodologies in Use*, Working group on Environmental Information and Outlooks, Report env/epoc/se(2001)2/final.

Osberg L. & Sharpe A. 1998, 'An Index of Economic Well-being for Canada' in *The State of Living Standards and Quality of Life in Canada*, Toronto: University of Toronto Press. See: http://www/csls.ca

Osberg L. & Sharpe A. 2000, 'Estimates of an Index of Economic Well-Being for OECD Countries', paper presented to the International Conference in Income and Wealth, Cracovia, 27 August–2 September.

Osberg L. & Sharpe A. 2002, 'An Index of Economic Well-Being for Selected Countries', *Review of Income and Wealth*, September.

Osberg L. & Sharpe A. 2003, 'Human Well-Being and Economic Well-Being: What Values are Implicit in Current Indices?', CSLS Research Report, 2003–04, August. See: http://www/csls.ca.

Perret B. 2002, *Indicateurs Sociaux, Etat des lieux et perspectives*, Report to the CERC, January.

Ponthieux S. 2003, 'Que faire du "Capital Social"?', Institut national de la statistique et des études économiques, Working Paper F0306, Paris.

Putnam R. 1995, 'Bowling Alone: America's Declining Social Capital', *Journal of Democracy*, 6(1), pp. 65–78.

Putnam R. 2000, *Bowling Alone, The Collapse and Revival of American Community*, Simon and Schuster, New York.

Sen A. 1987, *On Ethics and Economics*, Oxford Blackwell.

Sharpe A. 1999, *A Survey of Indicators of Economic and Social Well-Being*, paper prepared for Canadian Policy Research Networks, CSLS, Ottawa.

Sharpe A. 2003, *Literature Review of Frameworks for Macro-Indicators*, CSLS Research Report 2003–10, Ottawa.

Singh S. 2002, *Fermat's Last Theorem*, Fourth Estate, London.

United Nations 2003, *The Handbook of National Accounting: Integrated Environmental and Economic Accounting*. Online at: http://unstats.un.org/unsd/environment/seea2003.htm.

UNDP, *Human Development Report*, annually since 1990.

Vanoli A. 2002, *Une Histoire de la comptabilité nationale*, La Découverte, Paris.

Venetoulis J., Chazan D. & Gaudet C. 2004, 'Ecological Footprint of Nations, Sustainability Indicator Program', *Redefining Progress*, March. See: www.redefining.progress.org/publications/footprintnationas2004.pdf.

Wackernagel M. & Onisto L. (eds) 1997, *Ecological Footprint of Nations*. Online at: http://www.ecouncil.ac.cr/rio/focus/report/english/footprint.

Wackernagel M. & Rees W. 1995, *Our Ecological Footprint: Reducing Human Impact on the Earth*, New Society Publishers, Gabriola Island BC (the New Catalyst Bioregional Series).

Index

Printed in the United States
69979LV00001B/328